4

Home Favourites with Cookwares

4 廚具煮好菜

序

　　從專業院校導師轉型至烹飪中心導師，轉眼已數年了！教授的對象也跟從前的有所不同，由年輕學生變成家庭主婦及烹飪愛好者。現在與學生閒談交流煮食心得時，大部分環繞在家烹飪時遇到的種種難題，因此在編寫食譜時針對切身問題，讓讀者及學生得到解決及改善的方法。

　　現時，一般家庭的廚房面積有限，因此廚具設備的品種及款式不太多，通常以鑊、煲、電焗爐及電飯煲為主，在沒有特定廚具下如何煮出美味佳餚？又是讀者經常細問的。其實利用基本的廚具，如鑊、煲、電焗爐及電飯煲，也可作煎、炒、蒸、煮、炸，燜、燉及焗等多樣化的烹調方法。

　　今次推出的新書，以一般簡單廚具製作多款不同的家庭菜或宴客菜，達到最佳效果。希望這本新作能帶給讀者嶄新的烹飪靈感及構思，製作更多美味佳餚與至愛親朋共享。

　　藉此機會，特別感謝我的太太Jannies擔當廚務助理，在整個製作過程中悉心安排、協助及提供寶貴意見，以及圓方出版社製作團隊之合作，令此書順利面世。

廖教賢

目　錄
CONTENTS

歡聚。宴客菜

滋味。甜點

自製私房醬料

Home-made Sauce

喜愛入廚的你，必定愛親手弄醬料、煮醬汁，除了享受那份自家做的味道外，多做一些貯存於雪櫃，省卻每次弄醬的時間，實在是入廚的大智慧。

今次我介紹四款常用的醬汁，可靈活配搭不同的家常小菜或宴客菜式，而且貯存方法簡便，只需用熱水消毒玻璃容器，風乾，確保沒有水分及污染，倒入醬料後冷藏，一般可貯存約一個月或以上。

建議菜式：
茄汁煎中蝦 p.108，糖醋菊花魚 p.121
（＊或咕嚕肉、生炒排骨）

Sweet and Sour Sauce
糖醋汁

材料 Ingredients

白米醋	300 毫升
清水	300 毫升
片糖	2.5 片
OK 汁（酸甜調味醬）	5 湯匙
茄汁	180 克
喼汁	1 湯匙
鹽	1/3 茶匙
西檸（檸檬，後下）	1 個

300 ml white rice vinegar
300 ml water
2.5 pieces slab sugar
5 tbsp OK sauce
180 g ketchup
1 tbsp Worcestershire sauce
1/3 tsp salt
1 lemon (added last)

做法 Method

1. 西檸切片，備用。
2. 其餘材料用慢火煮溶，熄火，下鹽及檸檬片。
3. 待涼後，取出檸檬片棄掉，放入玻璃瓶盛起，冷藏貯存，使用時翻熱即可。

1. Slice the lemon and set aside.
2. Cook the other ingredients over low heat until they dissolve, turn off the heat, then add the salt and sliced lemon.
3. When it cools, discard the lemon, put the sauce into a jar and keep in a fridge. Heat up before using.

Liu's tips

- 糖醋汁煮成後，加入檸檬片待一會，令醬汁帶一陣檸檬清香味。建議涼透後即棄掉檸檬片，以免太久反而增加苦澀味。

- Add the sliced lemon into the cooked sweet and sour sauce, leave it for a while to give the sauce a light lemon fragrance. It is better to discard the lemon right after the sauce cools down. It will add a bitter taste to the sauce if the lemon is soaked for too long.

Fermented Black Bean Sauce
豆豉醬

建議菜式：
家鄉釀涼瓜 p.32、豉椒炒鱔球 p.98
（＊或豉汁蒸排骨、豉汁蒸魚）

材料 Ingredients

豆豉	300 克
薑	30 克
蒜茸	30 克
陳皮	1/3 個
砂糖	4 茶匙
生油	約半杯

300 g fermented black beans
30 g ginger
30 g finely chopped garlic
1/3 dried tangerine peel
4 tsp sugar
1/2 cup oil

做法 Method

1. 陳皮用水浸軟，刮去內瓤，切幼粒。

2. 薑切薄片，再切成幼粒。

3. 豆豉洗淨，瀝乾水分，剁碎。

4. 燒熱鑊，下生油約 2 湯匙，放入蒜茸及薑粒用慢火炒香，倒入其餘的生油燒熱，下豆豉碎及陳皮炒香，最後加砂糖煮溶。

5. 待涼後，放入密封玻璃瓶，冷藏貯存。

1. Soak the dried tangerine peel in water until soft, scrape off the pith and finely dice.
2. Thinly slice the ginger and then finely dice.
3. Rinse the fermented black beans, drain and finely chop.
4. Heat a wok, put in 2 tbsp of oil, add the garlic and ginger and stir-fry over low heat until aromatic. Pour in the rest oil, heat up, add the fermented black beans and dried tangerine peel, stir-fry until fragrant. Add the sugar and cook until it dissolves.
5. Put into a jar when it cools and keep in a fridge.

Liu's tips

- 盛入玻璃瓶後，必須有足夠的生油蓋過材料，若不足須加添足夠的份量。

- The amount of oil should be enough to cover the cooked ingredients in the jar. If it is short of oil, add more to the jar.

建議菜式：
杏香西檸雞 p.105
（＊西檸魚塊、西檸蒸烏頭）

Lemon Sauce
西檸汁

材料 Ingredients

西檸（檸檬）	1 個
濃縮檸檬果汁	200 克
白醋	150 毫升
清水	150 毫升
砂糖	180 克
吉士粉	30 克
鹽	1/3 茶匙（後下）

1 lemon
200 g concentrated lemon juice
150 ml white rice vinegar
150 ml water
180 g sugar
30 g custard powder
1/3 tsp salt (added last)

Liu's tips

- 以吉士粉勾芡，令西檸汁加添一抹果香味。

- The lemon sauce will have a hint of fruity flavour by thickening the sauce with custard powder.

做法 Method

1. 西檸磨去表面黃色的檸檬青，留用，切開檸檬榨汁留用。
2. 煮熱清水、白醋及濃縮檸檬果汁，加入砂糖煮溶。
3. 吉士粉用適量清水調溶，放入檸檬汁內勾芡，加入鹽、鮮檸汁及檸檬青拌勻。
4. 待涼後，放入密封玻璃瓶，冷藏貯存，使用時翻熱即可。

1. Zest over the yellow part of the lemon peel with a lemon zester. Cut open the lemon, squeeze out the juice and keep the juice.
2. Bring the water, white rice vinegar and concentrated lemon juice to the boil. Add the sugar and cook until it dissolves.
3. Dissolve the custard powder in some water, pour into the lemon juice mixture and stir until thick. Add the salt, fresh lemon juice and lemon zest, mix well.
4. When it cools, put into a sealed jar and then keep in a fridge. Heat up before using.

Western Style Sweet and Sour Sauce
西汁

建議菜式：
西汁烤肋排 p.80
（＊或中式牛柳、京都排骨）

調味料 Seasoning

茄汁	200 克
OK 汁（酸甜調味醬）	160 克
喼汁	1 湯匙
砂糖	200 克
罐裝牛尾湯	半罐（約 150 克）
花生醬	2 湯匙

200 g ketchup

160 g OK sauce

1 tbsp Worcestershire sauce

200 g sugar

1/2 canned oxtail soup (about 150 g)

2 tbsp peanut butter

香料水材料
Ingredients of spice water

清水	600 毫升
西芹	150 克
紅蘿蔔	150 克
番茄	1 個
洋葱	半個
紅尖椒	1 隻
香葉	6 片
草果	3 粒

600 ml water

150 g celery

150 g carrot

1 tomato

1/2 onion

1 red chilli

6 bay leaves

3 nutmegs

Liu's tips

- 以西芹、紅蘿蔔、香葉及草果熬成的香料水，是製作西汁的靈魂所在。
- 以花生醬作調味料，西汁的香味更濃郁。

- The spice water made by slowly cooking the celery, carrot, bay leaves and nutmegs is the soul of the sauce.
- The sauce flavoured with peanut butter has a stronger fragrance.

做法 Method

1. 香料水材料切碎；燒滾清水 600 毫升，加入材料用慢火煲約 40 分鐘，至餘下約 300 毫升香料水，棄掉湯渣，香料水留用。

2. 花生醬用香料水調溶，再倒回香料水內拌勻，加入牛尾湯煮開，下其他調味料，煮熱後待涼，放入玻璃瓶盛起，冷藏貯存，使用時翻熱即可。

1. Finely chop the ingredients of spice water. Bring 600 ml of water to the boil; add the ingredients and cook over low heat for about 40 minutes, or until the water reduces to about 300 ml. Discard the residual ingredients and keep the spice water.

2. Dissolve the peanut butter in some spice water, pour the mixture back to the spice water and mix well. Add the oxtail soup, bring to the boil, put in the other seasoning and heat up. When it cools, put into a jar and keep in a fridge. Heat up before using.

Moisturizing · Soups
滋潤。湯水

一個湯煲，
滋養身心的簡易小廚具，
讓身體，充滿能量！

A saucepot – the simple kitchen utensil –
can work out soups that nourish your body and mind.
Feeling great!

Coco-de-Mer, Chestnut and
Pork Shin Soup

海 底 椰 栗 子 煲 豬 腱

材料

乾海底椰	30 克
栗子	200 克
排骨	300 克
豬腱	400 克
薑	3 片（大）
蜜棗	4 粒
清水	約 8 杯

調味料

鹽	半茶匙

做法

1. 乾海底椰洗淨，用水浸泡約半小時。

2. 栗子去殼、去皮。

3. 豬腱及排骨洗淨，飛水備用。

4. 燒滾清水，放入所有材料用大火煲約半小時，再轉中火煲約 2 小時，灑入鹽調味即可。

Liu's tips

- 乾海底椰需用水浸泡後，才加入煲內煮，令其味道較易滲出。

- 栗子去除外殼後，放入微波爐用中火加熱一分鐘，可輕易去掉外皮。

海底椰栗子煲豬腱

Double-steamed Black-skinned Chicken Soup with
Agrocybe Aegerila Mushroom and Dried Scallop

茶樹菇瑤柱燉竹絲雞

材料

竹絲雞	半隻
乾茶樹菇	60 克
瑤柱	60 克
圓肉	20 克
瘦肉粒	80 克
薑	2 片
清水	4 杯

調味料

鹽	半茶匙

做法

1. 乾茶樹菇用清水浸約半小時，洗淨，去除細砂粒，備用。

2. 瑤柱用蓋面的清水浸軟，約需時 45 分鐘。

3. 竹絲雞及瘦肉粒洗淨，飛水，放入燉窩內，再一併放入茶樹菇、圓肉、薑片及瑤柱（連瑤柱水），加入清水後，加蓋密封，隔水燉約 3 小時，拌入鹽即可。

Liu's tips

- 瑤柱用清水浸至發脹及略軟身後，放入湯內燉製，令瑤柱的鮮味在燉湯內散發出來。

- 也可選用新鮮的茶樹菇，但不宜貯存太久，很易變壞，買回來後應立刻烹製。

Double-steamed Monkey-head Mushroom Soup
with Jinhua Ham and Winter Melon

金腿白玉燉猴頭菇

材料

猴頭菇	50 克
冬瓜	400 克
金華火腿	50 克
瘦肉	80 克
杞子	20 克
薑	8 片（小）
雞湯	1.5 杯
清水	1.5 杯

調味料

鹽	半茶匙

做法

1. 猴頭菇用清水浸約 1 小時，洗去表面雜質，擠去水分；杞子洗淨備用。

2. 冬瓜連皮切厚件；金華火腿及瘦肉切粒，飛水備用。

3. 雞湯及清水混和，煮滾成燉湯水，備用。

4. 將所有材料放入燉窩內，注入燉湯水加蓋密封，隔水燉約 2 小時 30 分鐘，最後拌入調味料即可享用。

Liu's tips

- 猴頭菇浸軟、洗淨後，必須擠去水分以去除菇之異味，否則影響湯水的味道。

- 若想宴客時賣相美觀，可將金華火腿切薄片，於冬瓜厚片中間切入一刀（不切斷為準），將火腿片夾於冬瓜片中，再排入燉窩內，成為「白玉火腿夾」。

白玉燉猴頭菇

Beef Thick Soup
with Egg and Spinach

蛋花菠菜牛肉羹

材料

免治牛肉	300 克
菠菜	300 克
鮮冬菇	6 朵
芫茜	1 棵
葱	1 棵
雞蛋	1 個（拂勻）
雞湯	2 杯
清水	2 杯

醃料

鹽	半茶匙
生粉	2 茶匙
清水	6 茶匙
胡椒粉	1/4 茶匙

調味料

鹽	1 茶匙
胡椒粉	1/4 茶匙
麻油	半茶匙

生粉獻

生粉	2 湯匙
清水	6 湯匙

做法

1. 免治牛肉與醃料拌勻待半小時，飛水備用。

2. 鮮冬菇洗淨，切幼粒，備用。

3. 菠菜洗淨，切碎，用滾水略灼至軟身，盛起，漂凍（用冷水略沖），壓去水分。

4. 芫茜及葱切成幼粒，備用。

5. 煲內倒入雞湯及清水煮滾，加入牛肉、鮮冬菇及菠菜碎煮熱，放入調味料及勾芡，徐徐倒入蛋液拌勻，灑上芫茜及葱粒享用。

Liu's tips

- 菠菜碎用滾水稍灼及漂冷水，以去掉草青味道，以及保持青綠色澤。

- 滾煮的免治牛肉要保持滑嫩口感，秘訣在於醃味時加入生粉拌勻。

蛋花菠菜牛肉羹

Snakehead Soup with Pears and Spareribs

雪梨排骨生魚湯

材料

雪梨	3 個
生魚	1 條（約 450 克）
排骨	300 克
薑	3 片
紅棗	8 粒（去核）
清水	約 8 杯

調味料

| 鹽 | 半茶匙 |

做法

1. 雪梨去皮，切成小件，去掉果心，用清水浸過果肉面，備用。

2. 排骨飛水，洗淨。

3. 燒滾水，放入雪梨、排骨及紅棗用大火煲滾。

4. 生魚洗淨，抹乾水分；燒熱鑊，下生油約 2 湯匙，放入薑片及生魚煎香至兩面呈微黃色，盛起，生魚及薑片放入煲湯魚袋內，轉放入大滾湯水中煲約半小時，即可轉中火再煲 1 小時 30 分鐘。

5. 飲用前先撇去表面油分，盛起湯渣，下調味料拌勻即可。

Liu's tips

- 雪梨需要去掉果心才煲湯，因為水果的果心加熱後帶酸味，影響湯水的味道。

雪梨排骨生魚湯

Double-steamed Pigeon Soup
with Conch and Dried Scallop

螺頭元貝燉乳鴿

材料		調味料	
乳鴿	2 隻	鹽	半茶匙
急凍螺頭	300 克		
瑤柱	60 克		
瘦肉粒	80 克		
南北杏	30 克		
南棗	4 粒		
薑	2 片		
清水	4 杯		

做法

1. 瑤柱用蓋面清水浸軟，約 45 分鐘。

2. 急凍螺頭切開，去掉腸臟，用生粉 2 茶匙及鹽 1 茶匙洗擦乾淨，以清水洗淨，切粗條。

3. 乳鴿及瘦肉粒洗淨後，與螺頭肉一起飛水，備用。

4. 將所有材料放入燉窩內，加入清水後，加蓋，隔水燉約 3 小時，拌入鹽調味即可。

Liu's tips

· 螺頭用生粉及鹽洗擦，去掉表面的污迹及冰雪氣味，以免影響湯水的鮮甜味道。

Tofu and Fish Head Thick Soup

滑豆腐魚雲羹

材料

大魚頭	1 個（約 800 克）
滑豆腐	1 件
瘦叉燒	160 克
菜心梗	8 條
甘筍	80 克
雞蛋	1 個（拂勻）
雞湯	2 杯
清水	2 杯
薑	2 片
芫茜	1 棵
葱	1 棵

調味料

鹽	1 茶匙
胡椒粉	1/4 茶匙
麻油	半茶匙

生粉獻

生粉	2 湯匙
清水	6 湯匙

做法

1. 大魚頭開邊，洗淨，排上薑片蒸熟，拆骨留肉，魚頭肉可用手搓散備用。

2. 菜心梗及甘筍切薄片，備用。

3. 叉燒切薄片；豆腐切幼粒。

4. 芫茜及葱切成幼粒，備用。

5. 煮熱雞湯及清水，放入所有材料煮滾，加入調味料拌勻，勾茨煮熱，拌入蛋液成蛋花狀，灑上芫茜及葱粒享用。

Liu's tips

- 大魚頭蒸熟後，除骨時要小心處理，須去掉魚頭所有小骨，以免進食時鯁喉。

- 所有材料宜切薄及切幼細，令湯羹的質感更嫩滑。

- 可選用免治豬肉代替叉燒片，視乎個人喜好而挑選。

滑豆腐魚雲羹

Coconut, Corn and Chicken Soup

椰子粟米煲雞湯

材料

去殼椰子	1 個
粟米	2 條
雪耳	1 朵
紅蘿蔔	1 條
光雞	半隻
排骨	300 克
南北杏	80 克
清水	約 10 杯

調味料

鹽	半茶匙

做法

1. 雪耳及南北杏用清水浸約半小時，雪耳剪去中央的黃色硬蒂。

2. 紅蘿蔔去皮，切小塊；粟米去衣，洗淨，斬件。

3. 光雞及排骨洗淨，飛水備用。

4. 椰子破開，倒出椰水留用，椰子肉切成小件，洗淨。

5. 將所有材料（雪耳除外）及椰子水放入煲內，用大火煲約半小時，加入雪耳轉中小火煲約 1 小時 30 分鐘，熄火，盛起湯渣，撇去油分，最後下鹽調味即可。

椰子粟米煲雞湯

Flavourful Memories · Home Dishes
回味。家常菜

小廚房的簡單廚具，
換一換用法，
也可蒸、炒、燜、煎、炸……
嘗試意想不到的美味體會。

By changing their original uses,
simple cookware in small kitchens
can also do other things –
steaming, stir-frying, braising, frying, deep-frying …
to give you an incredible experience of delicious food!

Stuffed Bitter Cucumber
家鄉釀涼瓜

材料

涼瓜	2 個

豬肉餡料

豬絞肉	400 克
蝦米	40 克
雞蛋	1 個

醃料

鹽	半茶匙
雞粉	半茶匙
生粉	2 湯匙

料頭

豆豉醬	1 湯匙
（參考 p.8）	
蒜茸	1 茶匙
薑粒	1 茶匙
葱粒	20 克

調味料

鹽	半茶匙
蠔油	1 湯匙
砂糖	1 茶匙
生抽	1 湯匙
老抽	1 茶匙
清水	1 杯

Liu's tips

- 先在涼瓜環內塗上少許乾生粉才釀入肉餡，在烹製過程中不易脫落。

- 用慢火煎至肉餡半熟，再倒入調味汁燜煮，可縮減燜煮時間，否則時間太長令涼瓜環散爛及顏色枯黃。

做法

1. 涼瓜橫切成約 2 厘米厚圓環狀，刮去瓜瓤中的籽及白瓤，用熱水焓約 3 分鐘，盛起，用水漂凍，吸乾水分備用。

2. 將餡料中的蝦米浸軟、切碎，與其他材料拌勻，加入醃料攪至起膠質，於涼瓜內環抹上生粉，釀入餡料。

3. 燒熱平底鑊，下生油 2 湯匙，放入已釀肉餡之涼瓜環，用慢火煎至兩面金黃，盛起。

4. 鑊內放入料頭爆香，加入調味料煮熱，下涼瓜環用慢火燜熟，最後用生粉水 2 湯匙將汁液勾芡即可。

家鄉釀涼瓜

Crisp Mud Carp Balls

脆皮鯪魚球

材料

豆腐卜	20 粒
鯪魚膠	300 克
陳皮	1 角
蝦米	30 克
葱	1 棵
芫茜	2 棵

鯪魚膠可在街市購買已調味的，一般鹹味已足夠，毋須加添鹽分，加入配料拌勻至起膠即可烹調。

炸豆腐卜時要控制油溫，別太大火，否則表面會過硬。

做法

1. 蝦米及陳皮用清水浸軟；葱及芫茜洗淨，將以上配料切幼粒，加入鯪魚膠內順一方向攪勻至起膠質。

2. 於豆腐卜中間剪開（不要剪斷），翻轉內裏成表面，釀入鯪魚膠餡料，壓緊封口。

3. 小煲內用中火燒熱炸油約 3 杯，放入魚膠豆卜炸熟至脆身（約 5 至 6 分鐘），盛起，瀝乾油分即成。

脆皮鯪魚球

Sichuan Spicy Boiled
Grass Carp Belly

水煮鯇魚腩

Liu's tips

- 魚肉用白醋醃後令肉質略結實,烹煮時不易散爛。
- 煮魚肉時湯料剛滾起即熄火,魚肉不會過熟。

材料

鯇魚腩肉	500 克
大豆芽	160 克
香芹	100 克
小青瓜	1 條
辣椒乾	80 克
花椒粒	3 湯匙
辣椒油	8 湯匙
薑	4 片
蒜子	8 粒
芫茜	3 棵

醃料

鹽	半茶匙
蛋白	1 個
生粉	2 茶匙
白醋	1 茶匙
花椒油	1 湯匙

調味料

川式香辣醬	2 湯匙
生抽	2 湯匙
老抽	1 湯匙
鹽	1.5 茶匙
雞粉	1 茶匙
砂糖	1 湯匙
花椒粒	1 湯匙
花椒油	1 湯匙
指天椒	2 隻
清水	6 杯

做法

1. 鯇魚肉洗淨、切片，下醃料拌勻，備用。

2. 青瓜切片；香芹切段、洗淨。

3. 平底鑊內下花椒油爆香薑、蒜及花椒粒，
   加入大豆芽炒香，放入青瓜及調味料煮
   約 3 分鐘，加入魚片及香芹煮熟，倒入
   鍋內備用。

4. 鑊內燒熱辣椒油，放入辣椒乾及花椒粒
   爆香，倒入鍋內，最後灑上芫茜即成。

水煮鯇魚腩

Steamed Egg with Shrimps

鮮蝦蒸水蛋

材料

雞蛋	3 個
鮮蝦	6 隻
葱花	1 茶匙

調味料

鹽	半茶匙
溫水	270 毫升

做法

1. 鮮蝦去頭除殼，留尾部，在蝦背剅開，挑腸，洗淨，飛水備用。

2. 雞蛋拂成蛋液，加入調味料拌勻，用密篩將蛋液水過濾，倒入窩碟內用大火蒸約 2 分鐘。

3. 打開蓋，加入鮮蝦，轉小火蒸約 4 至 5 分鐘，取出，加入適量熟油及生抽，灑上蔥花即可。

鮮蝦蒸水蛋

蒸水蛋時，蛋層不宜太深，也別直接用大火蒸熟，要先用大火再調至小火。

可在蒸製期間揭開鑊蓋，觀察雞蛋的成熟程度，以及減低鑊內溫度，以達到嫩滑的效果。

緊記蛋液必須與溫水拌勻，別倒入凍水；蛋液與溫水的比例是 1：1.5。

45

Preserved Meat Rice with Vegetable
and Dried Black Mushrooms

菜粒冬菇臘味飯

材料		豉油汁	
臘腸	2 條	生抽	3 湯匙
膶腸	1 條	老抽	2 茶匙
臘肉	1/3 條	砂糖	1 茶匙
芥蘭	120 克	生油	1 茶匙
冬菇	6 朵	麻油	半茶匙
白米	300 克	開水	3 湯匙
		胡椒粉	適量

做法

1. 臘腸、膶腸及臘肉飛水，切粒。

2. 冬菇浸軟，去蒂，切幼粒；芥蘭洗淨，切粒備用。

3. 豉油汁煮熱，備用。

4. 白米洗淨後，放入飯煲內加清水 300 毫升煲成白飯，煲約 15 分鐘及水分開始收乾時，將臘味粒及冬菇粒鋪在飯面，加蓋再煲約 5 分鐘，最後加入菜粒煲至飯全熟即可。

5. 進食時灑入豉油汁調味，拌勻即成。

Liu's tips

- 菜粒不宜太早放入烹煮，否則容易變黃及過熟，令賣相不佳。

- 煮豉油汁時可加入芫茜一棵，令豉油帶有芫茜的香氣。

Chicken and Tofu in Casserole

雞粒滑豆腐煲

材料

布包豆腐	2 件
雞髀肉	200 克
冬菇	4 朵
梅香鹹魚	1 小件
清雞湯	半杯

料頭

蒜茸、薑片及葱花
各適量

醃料

鹽	半茶匙
生粉	2 茶匙
清水	2 湯匙
麻油	半茶匙

調味料

鹽	1/3 茶匙
蠔油	2 茶匙
砂糖	半茶匙
老抽	1 茶匙
柱侯醬	1 茶匙
生粉	2 茶匙
清水	1 湯匙

- 豆腐粒放入熱鹽水內浸泡，目的是令豆腐表面結實一點，烹煮時不易散爛。

- 鹹魚必須煎香後才加入炒煮，令鹹魚香味更突出。

做法

1. 雞髀肉去皮、洗淨，切粒後下醃料拌勻。

2. 冬菇浸軟，去蒂，洗淨及飛水，沖冷水，切幼粒。

3. 梅香鹹魚用慢火煎香，切粒備用。

4. 豆腐切粒，放入熱水內浸過面，下鹽半茶匙拌勻。

5. 燒熱生油 1 湯匙，下雞粒炒熟，放入料頭爆香，加入清雞湯、冬菇粒及鹹魚粒炒勻。

6. 豆腐隔水後，放入鑊內用慢火燜 1 分鐘，最後放入調味料勾芡，煮滾即可轉放瓦煲享用。

雞粒滑豆腐煲

51

Braised Mandarin Fish with Radish

蘿蔔燜桂花魚

蘿蔔燜桂花魚

材料

桂花魚	1 條（約 500 克）
白蘿蔔	250 克
冬菇	8 朵
蒜子	10 粒
薑片、葱絲、紅椒絲各適量	

調味料

鹽	1 茶匙
砂糖	半茶匙
蠔油	2 湯匙
麻油	1 茶匙
胡椒粉	半茶匙
柱侯醬	1 湯匙
廚酒	1 湯匙
清水	2 杯

做法

1. 冬菇用清水浸至軟身（約 2 小時），去蒂，洗淨、飛水，每件切開兩塊，備用。
2. 桂花魚洗淨，抹乾水分，在魚背剞十字花紋。
3. 白蘿蔔去皮、切厚片，飛水約 5 分鐘備用。
4. 在桂花魚表皮塗上薄薄生粉，用中油溫炸至全熟，盛起，放入蒜子炸至金黃色，瀝去油分。
5. 鑊內放入薑片爆香，下調味料煮熱，加入蘿蔔、冬菇及炸蒜子，用中火燜約 20 分鐘。
6. 轉慢火，加入桂花魚再燜約 6 分鐘，盛起上碟，蘿蔔及冬菇伴邊，將湯汁勾芡扒在魚上，灑上葱絲及紅椒絲即可。

Liu's tips

- 在桂花魚表皮剞花紋，炸魚時魚肉較易熟透。
- 蒜子經油炸後，燜煮時不易散爛，可保持原粒形狀。

Shrimp Stuffed Squids with Pepper Salt

椒鹽百花釀鮮魷

材料

冷藏蝦仁	200 克
鮮魷魚	2 隻（約 600 克）
馬蹄肉	6 粒
椒鹽	半茶匙

料頭

蒜茸及紅椒粒各適量

調味料

鹽	1/3 茶匙
雞粉	1/3 茶匙
生粉	3 茶匙
麻油	半茶匙
胡椒粉	適量

Liu's tips

- 製作蝦膠最重要先用布吸乾水分，打出來的蝦膠才爽口；攪打蝦膠也必須順一個方向攪拌，才能令蝦膠產生黏力及彈性。

- 將鮮魷件飛水，目的是受熱定型，釀入蝦膠後不易脫落；在鮮魷上塗抹生粉後才釀入蝦膠，也有相同的效果。

- 蝦膠鮮魷不宜長時間炸，否則魷魚收縮太多令蝦肉較大，賣相不美觀，用中油溫炸約 3 分鐘即可。

椒鹽百花釀鮮魷

做法

1. 馬蹄肉剁碎；蝦仁解凍，洗淨、瀝乾，用乾布吸乾水分，放在砧板上用刀面拍蝦仁至散爛，再用刀背剁成蝦茸。

2. 蝦茸放入碗內，先用手攪擦至有黏力，加入調味料攪拌至起膠質，再用手搓搓至有彈力即成蝦膠，放入雪櫃冷藏備用。

3. 鮮魷魚剪開腹腔，取去內臟、去衣、除鬚，洗淨，在腹腔面用刀縱橫切成花紋，再切成小長方件（2厘米 x 4厘米），飛水及吸乾水分，在沒剞花紋面塗上薄薄生粉，釀入蝦膠。

4. 燒熱生油約2杯，放入蝦膠鮮魷用中油溫炸熟至金黃色，盛起。

5. 炸油盛起，下料頭爆香，放入蝦膠鮮魷炒勻，最後灑入椒鹽拌勻即可上碟。

56

Fried Meat Patties with Salted Fish and Lotus Root
鹹魚煎藕餅

材料		調味料	
免治豬肉	250 克	鹽	1/3 茶匙
鯪魚膠	80 克	蠔油	1 湯匙
梅香鹹魚	60 克	砂糖	1 茶匙
蓮藕	150 克	麻油	1 茶匙
葱花	2 茶匙	胡椒粉	適量
薑粒	2 茶匙	生粉	2 湯匙
		蛋液	半個

做法

1. 蓮藕去皮，切幼粒，洗淨後隔去水分。

2. 鹹魚用慢火煎香，切碎，倒入大碗內與免治豬肉、鯪魚膠及調味料攪勻至起膠質。

3. 最後放入蓮藕粒、葱花及薑粒拌勻，將肉料唧成小肉團，放於塗上薄薄一層油的碟上，略壓平成肉餅。

4. 燒熱鑊，下生油 3 湯匙，放入小肉餅用慢火煎香兩面至熟及金黃色即可。

Liu's tips

鹹魚煎藕餅

- 不可將未煎的鹹魚放入肉料內，必需先煎香及切碎粒，否則難以散發鹹魚的風味。
- 可於街市購買已調味的鯪魚膠。

Baked Mackerel Steaks in Teriyaki Sauce

燒汁煎焗鮫魚扒

材料

鮫魚扒	2 件
洋葱絲	半個
日式照燒汁	3 湯匙
沙律醬	適量

醃魚汁 A

生抽	3 湯匙
老抽	2 湯匙
鹽	1 茶匙
雞粉	2 茶匙
砂糖	2 茶匙
花雕酒	1 湯匙
胡椒粉	半茶匙
香葉	4 片
清水	1 杯

醃魚汁 B

香芹	150 克
甘筍	150 克
乾葱	4 粒
洋葱	半個
薑	10 克
芫茜	3 棵
清水	1 杯
青檸	2 個 （榨汁）

做法

1. 醃魚汁 A 用慢火煮溶調味料，待涼。

2. 醃魚汁 B 用攪拌機攪碎，隔渣留汁。將醃魚汁 A 及 B 混合，放入已洗淨的鮫魚扒醃浸約 4 小時。

3. 魚扒抹乾水分，燒熱鑊下油用中火煎香魚扒兩面，盛起。

4. 預熱焗爐240℃，焗盤鋪上錫紙及洋葱絲，鮫魚扒塗上照燒汁，放在洋葱絲上焗約 5 分鐘，取出，再塗照燒汁反轉焗 5 分鐘至熟，享用時蘸上沙律醬。

Liu's tips

- 鮫魚扒不宜切得太薄（約 2 厘米厚度較適合），否則受熱後收縮很易散爛。

- 若想魚扒表面呈脆香效果，先在魚扒兩面塗上少許麵粉，煎至金黃色後再烤焗。

燒汁煎焗鮫魚扒

Deep-fried Spareribs
and Pumpkin in Apricot Sauce

金瓜黃梅骨

材料

一字排	300 克
日本南瓜	300 克
粟米片	50 克

醃料

鹽	半茶匙
雞粉	半茶匙
生粉	2 茶匙
清水	2 湯匙
吉士粉	1 茶匙
麵粉	2 湯匙（後下）

黃梅醬調味料

黃梅果醬	6 湯匙
泰國青檸	2 個
砂糖	1 湯匙

Liu's tips

- 日本南瓜皮較薄，可以不用削去皮，但如選用中國南瓜則需要削去表皮，因中國南瓜皮厚，以免影響質感。

做法

1. 一字排斬成長骨排形，洗淨，瀝乾水分，用醃料拌勻待半小時。

2. 日本南瓜洗淨，連皮切成骨排形；青檸磨去表皮之檸檬皮青，留用，並榨出青檸汁約 1 湯匙。

3. 將黃梅果醬、青檸汁和砂糖用慢火煮熱，加入檸檬皮青攪勻成黃梅醬汁。

4. 鑊內放入生油約 2 杯燒熱，排骨用麵粉拌勻，放入油鑊內用中油溫炸熟，盛起，瀝乾油分，再放入南瓜炸脆。

5. 煮熱黃梅醬汁，放入排骨及南瓜拌勻，上碟，灑上略壓碎之粟米片即可。

金瓜黃梅骨

Steamed Meat Patty
with Salted Egg and Water Chestnut

鹹蛋馬蹄蒸肉餅

材料

梅頭肉	300 克
鹹蛋	1 個
馬蹄肉	6 粒
冬菇	4 朵

調味料

鹽	1/3 茶匙
蠔油	1 湯匙
生粉	2 茶匙
清水	2 湯匙

Liu's tips

- 也可購買用攪肉機攪碎的豬肉，但肉質比較鬆散，彈力略差。

- 肉團調味後不可過分攪撻，否則蒸熟後令肉餅收縮結實，口感不好。

- 肉團鋪平碟上時不要太厚，否則難以蒸熟及質感欠佳。

- 馬蹄肉放入密實袋後拍散，較容易處理。

- 壓蛋黃時，先用蛋白塗抹在菜刀面上，以防蛋黃黏緊。

鹹蛋馬蹄蒸肉餅

做法

1. 梅頭肉洗淨，切小件、剁碎；馬蹄肉拍散，切碎粒；冬菇浸軟，去蒂，切幼粒。

2. 鹹蛋分開蛋白及蛋黃，蛋黃用菜刀壓平成圓形，蛋白留作調味之用。

3. 梅頭肉碎放入大碗內，加入調味料及鹹蛋白，順方向攪拌至肉碎起膠質及帶黏性。

4. 用手將肉團撻 5 至 6 次，放在碟中壓平，在中央刮開小洞，放上鹹蛋黃。

5. 放入鑊內隔水蒸約 10 分鐘，取出，灑上葱粒即可享用。

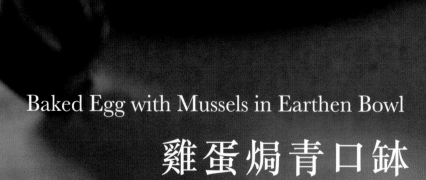

Baked Egg with Mussels in Earthen Bowl

雞蛋焗青口鉢

材料

急凍青口肉	12 粒
豬肉碎	50 克
肥肉粒	30 克
陳皮	1 角
薑米	1 茶匙
蒜茸	1 茶匙
葱花	1 茶匙
蛋液	1 個
油條	1/4 條

調味料

雞蛋	3 個
鹽	半茶匙
雞粉	1/3 茶匙
清水	（與雞蛋比例是 1:1）
麻油	半茶匙
胡椒粉	1/3 茶匙

青口鉢可預先蒸熟，待涼後放入雪櫃貯存，烹製時取出蒸熟，塗上蛋液烘焗即可，省卻時間。

如選用的瓦鉢較小，材料拌勻後倒入鉢內宜八成滿，份量不要太滿，否則蒸熟後空間不足塗上蛋液，令烘焗後膨脹滿瀉。

雞蛋焗青口鉢

做法

1. 青口肉洗淨、飛水，切粗粒後放入瓦鉢內。

2. 陳皮用水浸軟，刮去瓤，切碎；油條切片。

3. 豬肉碎及肥肉粒放入鑊內，下薑米、蒜茸及陳皮同炒香，放入瓦鉢。

4. 調味料拌勻，加入葱花後倒入瓦鉢內，排上油條，放入鑊內蒸約12分鐘至熟，待涼，用乾布或廚房紙吸去表面水分。

5. 預熱焗爐240℃，在青口鉢表面塗上蛋液，放入焗爐焗約8分鐘，取出，塗上蛋液再焗約5分鐘至金黃色即可。

Steamed Pork Belly with
Preserved Flowering Cabbage

梅 菜 扣 肉

材料

五花腩肉	500 克
梅菜	300 克

料頭

蒜茸	1 茶匙
乾葱茸	1 茶匙

調味料

柱侯醬	2 湯匙
南乳	1 湯匙
蠔油	2 湯匙
鹽	半茶匙
砂糖	1 湯匙
生抽	1 湯匙
老抽	1 茶匙
玫瑰露酒	2 茶匙
清水	1 杯

Liu's tips

- 梅菜浸軟後，清洗時要將菜葉張開，洗去幼砂粒，以免藏在菜葉中。

- 焓五花腩時，熱水不可調大火，以免腩肉過腍，切厚片時皮層很易鬆散。

做法

1. 梅菜撕開成小棵，用清水浸軟約 1 小時去掉鹹味，洗掉幼砂粒，擠乾水分，切粒。

2. 五花腩用熱水慢火焓約 45 分鐘，盛起，洗去表面油脂，在表皮塗抹老抽上色。熱鑊下生油少許將五花腩表皮煎至有色，備用。

3. 燒熱鑊，下梅菜粒乾炒至香，下砂糖約 2 湯匙炒香至微黃，下生油 2 湯匙拌勻，備用。

4. 五花腩切厚片；燒熱油鑊，下料頭起鑊，放入五花腩及調味料爆香後，將腩肉排列大碗內，鋪入梅菜粒蓋在腩肉上，淋上調味汁。

5. 將整碗扣肉隔水蒸約 2 小時，取出，倒起醬汁勾芡，扒在扣肉上便可。

梅菜扣肉

73

Baked Chicken Wings
with Sesame Seeds

芝麻雞中翼

材料

雞中翼	10	隻
白芝麻	80	克
雞蛋白	1	個
粟粉	2	茶匙

醃料

生抽	1	湯匙
老抽	2	茶匙
麻油	1	茶匙

滷水料

八角	6	粒
花椒粒	1	茶匙
桂皮	1	片
沙薑	10	片
香葉	10	片
薑	4	片
葱	2	棵
清水	5	杯

調味料

鹽	3	茶匙
雞粉	2	茶匙
砂糖	3	茶匙
花雕酒	2	湯匙

Liu's tips

- 雞中翼浸泡滷水時毋須開火，否則雞翼過熟令肉質太腍。

- 雞翼風乾的方法：可用電風扇調至微風幫助吹乾，減少風乾時間。

做法

1. 滷水料放入小煲內，用慢火煮約 1 小時，放入調味料煮溶。

2. 雞中翼解凍，下醃料醃約 1 小時，放入熱滷水汁內，熄火浸焗約 15 分鐘，盛起，放於焗爐架上，瀝乾水分。

3. 雞蛋白及粟粉拌勻，塗抹雞中翼表皮，灑上白芝麻，風乾約 2 小時。

4. 預熱焗爐 240 ℃，放入雞翼焗約 10 分鐘，取出，反轉另一面再焗 10 分鐘至金黃色即可。

芝麻雞中翼

Salted Fish, Abalone and Chicken Rice

鹹魚鮮鮑雞粒飯

材料

鮮鮑魚	6 隻
雞髀扒	1 件
馬友鹹魚	1 小件
（約 40 克）	
冬菇	4 朵
薑絲	20 克
葱花	20 克
白米	300 克

醃料

鹽	半茶匙
蠔油	1 湯匙
生粉	3 茶匙
清水	3 湯匙

豉油汁

生抽	3 湯匙
老抽	2 茶匙
砂糖	1 茶匙
生油	1 茶匙
麻油	半茶匙
開水	3 湯匙
胡椒粉	適量

Liu's tips

- 鮮鮑魚去掉內臟後，用生粉拌勻，放在箸箕內不斷攪擦，擦掉表面污漬，最後用清水沖洗乾淨。

78

做法

1. 鮮鮑魚去掉外殼，清除內臟，洗淨，瀝乾水分，切粗粒。

2. 雞髀扒去皮、切粒，混和醃料後待約 15 分鐘，再與鮮鮑魚粒拌勻。

3. 冬菇用水浸軟，去蒂，切粒；豉油汁煮熱，備用。

4. 白米洗淨後，放入飯煲內加清水 300 毫升煲成白飯，煲約 15 分鐘及開始水分收乾時，加入雞粒、鮑魚粒及冬菇粒鋪在飯面，再放上鹹魚件，並灑上薑絲，加蓋煲至飯熟，焗約 5 分鐘。

5. 進食時灑上葱花，加入豉油汁拌勻調味即成。

鹹魚鮮鮑雞粒飯

Roasted Spareribs in
Western Style Sweet and Sour Sauce
西汁烤肋排

西汁烤肋排

材料

肋排	600 克
西汁	6 湯匙（參考 p.10）
香草粉	1 湯匙

醃料

鹽	1 茶匙
雞粉	半茶匙
砂糖	1 茶匙
生粉	3 茶匙
雞蛋黃	1 個

做法

1. 肋排斬件約 6 厘米長度，洗淨後醃約 1 小時，再用西汁 3 湯匙拌勻。
2. 預熱焗爐 200℃，肋排鋪在已墊錫紙的焗盤上，焗約 10 分鐘，取出，反轉另一面再焗 10 分鐘。
3. 肋排再塗上西汁及灑上香草粉，將焗爐溫度調高至 240℃，放入肋排每面烤焗 5 分鐘即成，進食時蘸上西汁。

Liu's tips

- 西汁可預先調較妥當貯存於雪櫃，此醬汁可烹製中式牛柳、京都肉排或炸豬扒等菜餚，不妨預備在家隨時使用。

Deep-fried Chicken and Lotus Root with Red Fermented Beancurd

南乳藕片碎炸雞

Liu's tips

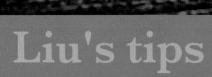

蓮藕宜選用外形較瘦及呈長身的較為合適。

蓮藕必需切成薄片（約 2 至 3 毫米），如切得太厚難以炸至脆身。炸製時注意油溫不能太高，約 120℃ 炸至脆身即可。

材料

冰鮮光雞	半隻
蓮藕	200 克
花雕酒	1 茶匙

料頭

蒜茸、洋葱粒、葱粒、紅椒粒各 1 茶匙

醃料

南乳	2 大磚
雞粉	半茶匙
蠔油	1 湯匙
砂糖	1 湯匙
蒜茸	1 湯匙
薑粒	1 湯匙
生粉	2 茶匙
低筋麵粉	3 湯匙（後下）

做法

1. 蓮藕去皮，洗淨，切薄片；取醃料中南乳半磚及砂糖 1 湯匙，與清水 1 杯 攪溶，浸泡蓮藕片約半小時，備用。

2. 光雞斬件，洗淨，瀝乾水分，下醃料拌勻醃約半小時。

3. 燒熱油，將雞件蘸上麵粉，放入熱油炸熟至金黃色，盛起。

4. 蓮藕片隔去水分，用乾布吸乾水分，放入熱油內炸至金黃色，盛起，瀝乾油分。

5. 燒熱油，下料頭爆香，放入炸雞件拌炒，最後灑上花雕酒炒勻，上碟，以炸蓮藕片伴碟即成。

南乳藕片碎炸雞

Eggplant with Pork and
Salted Fish in Casserole

魚香茄子煲

- 茄子削皮及切段後，若非即時烹煮，建議放入稀釋的鹽水內浸泡，以免氧化令表面變成褐色。

- 鹹魚可選用梅香鹹魚，香氣濃，較有風味。

材料

茄子	300 克
豬肉碎	160 克
冬菇	3 朵
梅香鹹魚	1 小件（約 20 克）
清水	1 杯

料頭

蒜茸、薑粒及蔥花各少許

醃料

鹽	1/3 茶匙
生粉	2 茶匙
麻油	1 茶匙

調味料

鹽	1/3 茶匙
砂糖	半茶匙
蠔油	2 茶匙
豆瓣醬	1 茶匙
老抽	1 茶匙
清水	1 杯

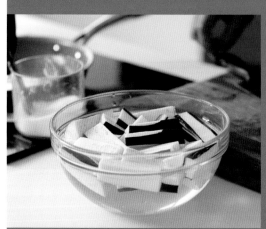

做法

1. 鑊內下生油 2 湯匙燒熱，用慢火煎香鹹魚件，切幼粒備用。

2. 豬肉碎與醃料拌勻；燒熱鑊後下生油 1 湯匙，放入肉碎炒熟，盛起。

3. 冬菇浸軟，去蒂，切幼粒；茄子洗淨，相間地削皮、切段，鑊內下生油半杯燒熱，放入茄子炸至軟身及金黃色，隔去油分。

4. 燒熱油，放入料頭用中火爆香，加入所有材料用中火炒勻，拌入調味料，轉慢火燜至茄子腍身，用生粉水勾芡煮熟，轉放入瓦煲享用。

魚香茄子煲

Stir-fried Sea Cucumber with
Zucchini and Bell Peppers
雙脆炒海參

材料

急凍海參	200 克（已浸發）
翠玉瓜	150 克
木耳	40 克
三色甜椒	100 克

滾煨料

鹽	1 茶匙
麻油	1 茶匙
雞粉	1 茶匙
清水	1 杯
廚酒	1 茶匙

料頭

蒜茸、薑片、甘筍片各適量

芡汁

鹽	1/3 茶匙
蠔油	1 湯匙
麻油	半茶匙
生粉	2 茶匙
清水	1 湯匙

做法

1. 急凍海參及木耳用清水浸軟,放入熱水
   內,下薑汁酒2湯匙滾透去異味,盛起,
   用冷水沖至涼。

2. 海參及木耳切條,放入已加入滾煨料的清
   水內煮一會,連湯汁盛起浸泡備用。

3. 翠玉瓜及三色甜椒洗淨、切條;燒熱鑊下
   生油1湯匙,放入翠玉瓜用中火煽炒,放
   入薑汁酒1湯匙及鹽半茶匙,炒約2分鐘,
   加入三色甜椒再炒約1分鐘,隔水盛起。

4. 海參及木耳隔去滾煨湯汁,備用。

5. 鑊燒熱,下生油1湯匙,放入料頭爆香,
   加入所有材料用中火炒勻,倒入已拌勻的
   芡汁,用大火快炒至熱,上碟享用。

薑汁酒製法:薑肉300克、冷開水300毫升(薑與水的比例是一比一),全部材料放入攪拌機搗碎,隔渣留汁,加入廚酒50毫升混合即成薑汁酒,能辟除材料的腥膻異味,可貯存雪櫃備用。

雙脆炒海參

Minced Beef Rice with
Dried Octopus and Water Chestnut

鱆魚馬蹄牛崧飯

材料

免治牛肉	250 克
馬蹄肉	6 粒
冬菇	5 朵
乾鱆魚	200 克
薑粒	20 克
葱花	20 克
陳皮末	10 克

醃料

生抽	1 湯匙
蠔油	1 湯匙
生粉	3 茶匙
清水	4 湯匙
胡椒粉	1/4 茶匙
麻油	1 茶匙

豉油汁

生抽	3 湯匙
老抽	2 茶匙
砂糖	1 茶匙
生油	1 茶匙
麻油	半茶匙
開水	3 湯匙
胡椒粉	適量

做法

1. 乾鱆魚浸水至軟身，切粒備用。

2. 冬菇浸軟，去蒂，切粒；馬蹄肉用刀拍散，剁成碎粒；免治牛肉與醃料拌匀，加入冬菇粒、馬蹄肉、陳皮末及薑粒攪匀至有膠質。

3. 豉油汁煮熱，備用。

4. 白米洗淨後，與鱆魚粒一起放入飯煲內，放清水300毫升煲約 15 分鐘，見飯水開始收乾時，加入牛肉碎鋪平在飯面，加蓋煲熟，焗約 5 分鐘即可。

5. 進食時灑上葱花，加入豉油汁調味，拌匀享用。

乾鱆魚浸軟後才容易切成粒，與白米及水一起煲煮，令鱆魚的香味滲透白飯內。

鱆魚馬蹄牛崧飯

Happy Gathering · Banquet Dishes
歡聚。宴客菜

基本的家用廚具，
使精巧的菜餚成為飯聚的指定菜式，
在家聚首歡慶，輕鬆自在！

Want to put exquisite dishes on the menu
to entertain your friends at home?
Basic household cookware can do the job.
Enjoy an amazing and relaxing party time!

Stir-fried Mixed Vegetables
in Taro Nest

鵲巢炒素丁

材料

鮮冬菇	8 朵
蘑菇及鮮草菇	各 8 粒
西芹	150 克
甘筍	80 克
素肉	120 克
方包	1 片
芋頭	300 克

調味料

鹽	1/3 茶匙
蠔油	1 湯匙
生粉	2 茶匙
清水	2 湯匙

- 難以脫模。
則炸芋絲會黏着鵲巢兜，
面，取出後排上芋絲，否
再放入中油溫的熱油浸過
內消毒，用布抹至乾爽，
鵲巢兜使用前先放入熱水

- 處理。
化芋絲，排成鵲巢時容易
芋絲先用鹽輕拌，能夠軟

- 煩惱。
去皮後才清洗，不會有此
容易令人手部痕癢，建議
芋頭表皮含草酸，處理後

- 取出待涼。
120℃焗約5分鐘呈金黃色，
法處理，放在焗盤上，用
麵包粒也可使用烘焗的方

鵲巢做法

1. 芋頭切薄片，再切成幼絲，用鹽 2 茶匙拌勻醃約 8 分鐘，洗去鹽分，瀝水後用乾布盡量吸乾水分。

2. 鑊內倒入油至半鑊，用中火燒熱，放入鵲巢兜 2 個用熱油蓋面，盛起備用。

3. 芋絲與麵粉 2 湯匙拌勻，先放入數條芋絲在鵲巢兜底部，再將 3 至 4 條芋絲結成一圈，一層層疊排於鵲巢兜邊緣，砌成花籃形狀，最後再放數條芋絲在底部，以防甩脫。

4. 用另一個鵲巢兜放在盛有芋絲的兜內，輕輕壓緊，放入熱油鑊內用中油溫炸至金黃色（或可脫模），脫模後多炸一會，盛起即可。

做法

1. 鮮冬菇切十字形，切成四小件；蘑菇及鮮草菇切半，用薑汁酒滾透，盛起，用冷水沖透，備用。

2. 素肉、西芹及甘筍切粒，備用。

3. 方包去掉邊皮，切丁方粒，用中油溫炸脆，呈金黃色即盛起，用吸油紙吸去油分，待涼。

4. 鑊內注入清水 1 杯、鹽半茶匙及油 1 茶匙煮滾，放入西芹及甘筍粒滾煨約 1 分鐘，再下菇料及素肉粒煮熱，隔水備用。

5. 鑊燒熱後，下油 1 茶匙，加入所有材料炒勻，調味料混和後拌入炒勻，放入鵲巢內，最後灑上脆麵包粒即可享用。

鵲巢炒素丁

Stir-fried Japanese Eel in Chilli Bean Sauce

豉椒炒鱔球

材料

白鱔肉	500 克
青甜椒	半隻
紅尖椒	1 隻
洋葱	半個

料頭

蒜茸	1 茶匙
豆豉醬	1 湯匙
（參考 p.8）	
葱粒	20 克

醃料

鹽	半茶匙
生粉	3 茶匙
胡椒粉及麻油各適量	
清水	3 茶匙

調味料

鹽	1/3 茶匙
砂糖	半茶匙
蠔油	1 湯匙
老抽	2 茶匙
生粉	2 茶匙
清水	2 湯匙
胡椒粉及麻油各適量	

做法

1. 白鱔肉洗淨，先直切後橫切剁成花紋，再切成 3 厘米 x 6 厘米長方件，即成鱔球，下醃料拌勻，備用。

2. 青甜椒、紅尖椒及洋葱洗淨，切角。

3. 燒熱鑊，用中油溫將白鱔球泡油至熟，盛起。

4. 鑊內放入料頭爆香，加入青、紅椒及洋葱用慢火炒熟，轉中火，加入鱔球略炒，拌入已混和的調味料爆炒拌勻，上碟享用。

Liu's tips

- 購買白鱔時，可請代勞起骨留肉，若表面的潺迹未徹底去除，可用食鹽 2 茶匙塗在鱔肉表皮待約 1 分鐘；或用約 75℃暖水，放入白鱔略浸 1 分鐘，令表面淡白色的潺迹浮現，用刀刮去後洗淨即可，以上兩種方法可取其一。

豉椒炒鱔球

Crab Satay with
Mung Bean Vermicelli
in Casserole

沙爹粉絲蟹煲

材料

肉蟹	1 隻（約 600 克）
粉絲	1 小紮
薑片	100 克
葱段	80 克
香芹	100 克
椰奶	80 毫升

料頭

蒜茸	1 茶匙
紅椒粒	1 湯匙

調味料

鹽	1 茶匙
砂糖	1 茶匙
雞粉	半茶匙
沙爹醬	2 湯匙
清水	1.5 杯

做法

1. 粉絲用清水浸軟，隔去水分。沙爹醬 1 湯匙及清水半杯煮熱，下粉絲浸泡備用。

2. 在肉蟹腹中斬入，揭開蟹蓋，去掉內臟，洗淨及斬件，瀝乾水分，灑上生粉 1 湯匙拌勻。

3. 燒熱生油 2 湯匙，放入蟹件用中火煎香，再下薑片及料頭爆香，加入調味料煮熱，加蓋，轉中慢火待約 6 分鐘。

4. 打開鑊蓋，拌入葱段及香芹，轉慢火加蓋待約 2 分鐘，熄火，最後倒入椰奶拌勻。

5. 粉絲隔水後放入瓦煲內，排上沙爹蟹件即成。

- 粉絲不宜太早拌入調味料與肉蟹一起烹煮，容易發脹及過於軟腍，只要於完成後將粉絲放於煲底，令粉絲易於吸收汁液精華，使其入味即可。

沙爹粉絲蟹煲

Seafood Rolls with Mashed Taro

荔茸海鮮卷

Liu's tips

- 半熟澄麵糰加入芋茸搓勻後，宜待涼後才加入固體油搓成荔茸，否則熱度容易令固體油溶化，令荔茸鬆散。

材料

急凍帶子	12 粒
雪藏蝦仁	12 隻
春卷皮	10 片

荔茸材料

芋頭	300 克
澄麵	40 克
植物固體油	30 克

荔茸調味料

鹽	半茶匙
雞粉	半茶匙
五香粉	半茶匙
麻油	半茶匙

海鮮調味料

鹽	半茶匙
生粉	1 茶匙
麻油	1 茶匙

麵粉漿

麵粉	2 湯匙
清水	4 湯匙

做法

1. 芋頭切件，蒸至軟身，壓成芋茸備用。

2. 澄麵用滾水 40 毫升拌勻，搓成半熟麵糰，與芋茸及調味料拌勻，最後加入植物固體油搓至光滑。

3. 帶子及蝦仁洗淨，用乾布吸乾水分，拌入海鮮調味料醃約半小時，切粗粒，灼熟後瀝乾水分，加入荔茸內拌勻成餡料。

4. 麵粉漿混合拌成黏漿，備用。

5. 春卷皮鋪平，放入荔茸餡包成長方春卷形，用麵粉漿黏住封口，放入熱油內用中油溫炸脆至金黃色即可。

荔茸海鮮卷

Deep-fried Almond Chicken in Lemon Sauce
杏香西檸雞

材料

雞髀扒	1 件
杏仁片	200 克
西檸汁	100 毫升
（參考 p.9）	
檸檬	4 片
生粉	2 茶匙
（用水拌勻）	

醃料

鹽	半茶匙
雞蛋黃	1 個
生粉	2 茶匙

蛋漿料

雞蛋	1 個（拂勻）
生粉	4 湯匙

杏香西檸雞

做法

1. 雞髀扒洗淨，瀝乾水分，加入醃料拌勻醃約半
   小時。

2. 蛋漿料混合攪拌，倒入雞髀扒內拌勻，表面再
   滾上杏仁片。

3. 鑊內注入生油半鑊，燒熱至中油溫（約
   140℃），放入雞扒炸約 1 分鐘，轉中慢火（約
   100℃）再炸約 3 分鐘至熟，最後轉回中油溫
   炸脆表面及呈金黃色，盛起，瀝去油分，切成
   小塊上碟。

4. 煮熱西檸汁和檸檬片，用生粉水 2 茶匙勾芡煮
   熟，扒在雞件上即成。

Fried Prawns in Sweet and Sour Sauce

茄汁煎中蝦

材料

鮮中蝦	12 隻
洋葱	80 克

料頭

蒜茸、薑片及葱粒各適量

調味汁

鹽	半茶匙
雞粉	1/3 茶匙
砂糖	2 茶匙
茄汁	3 湯匙
糖醋汁	2 湯匙（參考 p.7）
生粉	1 茶匙

做法

1. 鮮中蝦剪去鬚爪，洗淨，瀝乾水分；洋葱切粗絲。

2. 燒熱鑊，下生油 2 湯匙，用中火煎中蝦至兩面金黃及熟透，盛起。

3. 放入洋葱絲炒香，加入料頭爆香，下中蝦用中火炒勻，倒入已拌勻的調味汁，用大火炒勻即可享用。

• 修剪蝦鬚及爪時，可在蝦背剪開，除可挑去蝦腸外，也令中蝦容易熟透及入味。

茄汁煎中蝦

Braised Dongpo Pork with
Winter Melon

東坡白玉

Liu's tips

- 因不想太肥膩，所以切掉五花腩肥肉部分，改用冬瓜件代替，口感清爽。

- 切冬瓜時，適宜將面層2厘米留用，因面層質地較實，烹製時不易散爛。

材料

冬瓜	600	克
五花腩肉	600	克
韭菜	80	克

料頭

薑片	120	克
葱段	100	克
八角	5	粒
陳皮	1	片

調味料

鹽	1	茶匙
蠔油	2	湯匙
柱侯醬	1	湯匙
生抽	2	湯匙
老抽	1	湯匙
糖	1	湯匙
花雕酒	100	毫升
清水	3	杯

做法

1. 冬瓜切成 4 厘米丁方形（約 2 厘米厚），用熱水焓約 5 分鐘，盛起，沖冷水備用。

2. 五花腩肉連皮切去表面脂肪層，餘下較瘦的肉層約 2 厘米厚，洗淨。

3. 煮熱調味料，倒入電飯煲，加入料頭及五花腩煲約 1.5 小時，取出待涼，切成 4 厘米丁方形，汁液留用。

4. 韮菜用熱水浸軟，沖冷水備用。

5. 將冬瓜方件放於五花腩肉上，用韮菜打成十字結固定，放入窩碟內，注入剩餘汁液蓋過面，蒸約 15 分鐘至冬瓜腍身，放於碟上，用汁液勾芡扒上即可。

東坡白玉

Steamed Dried Scallops
and Deep-fried Garlic
with Chinese Lettuce

生財蒜子瑤柱甫

材料

原粒瑤柱	12 粒
蒜子	80 克
唐生菜	300 克
薑	1 片

芡汁調味料

蠔油	1 湯匙
老抽	1 茶匙
生粉	2 湯匙

Liu's tips

- 乾瑤柱浸泡前最宜先摘去圓邊旁的小硬塊，否則蒸腍後會有點韌。
- 原粒乾瑤柱必須先用冷水浸泡，令瑤柱吸收水分發脹後才蒸，否則難以使瑤柱完全脹大及軟腍。

做法

1. 原粒乾瑤柱放入小窩碟內，倒入清水約 1 杯浸過表面，浸泡約 2 小時。

2. 蒜子去衣、洗淨；鑊內下油 1 湯匙，放入蒜子用小火煎至金黃色，排入已浸泡之瑤柱上，再加薑 1 片，隔水蒸約 1 小時。

3. 唐生菜洗淨，灼熟後盛起，隔去水分。

4. 將瑤柱窩碟取出，徐徐倒出瑤柱水留用，放入唐生菜在瑤柱窩碟內略按，用另一大窩碟覆蓋壓緊，反轉後盛着瑤柱的小窩碟在上、大窩碟在底部。

5. 鑊內放入瑤柱水，下調味料煮成芡汁，拿開小窩碟，將瑤柱芡汁扒在瑤柱甫上即成。

115

Baked Pork Hock in
Chinese Marinade

滷焗豬手

材料

豬手	2 隻

料頭

乾葱頭	10 粒
蒜子	6 粒
薑	4 片

滷水料

八角	6 粒
陳皮	2 角
沙薑片	8 片
香葉	8 片

調味料

南乳	2 磚
鹽	2 茶匙
雞粉	2 茶匙
蠔油	3 湯匙
冰糖	30 克
花雕酒	3 湯匙
黑胡椒粒	2 茶匙
柱侯醬	2 湯匙
清水	8 杯

做法

1. 豬手斬件，飛水、洗淨，與滷水料放於電飯煲內。

2. 燒熱鑊，下生油 2 湯匙，爆香料頭，放入調味料煮熱，倒入電飯煲內燜約 1 小時 30 分鐘，熄火後焗半小時，取出豬手，表皮向上放於焗盤。

3. 焗爐先預熱 240℃ 待 10 分鐘，放入豬手焗約 15 分鐘，上碟即成。

Liu's tips

- 豬手放入焗爐時，宜抹去表皮水分，而且皮層向上，能夠將豬皮焗至香脆。

滷焗豬手

Cheese Baked Tiger Prawns with Bacon

煙肉芝士焗大花蝦

材料

大花蝦	6 隻
洋葱	1 個
煙肉	3 片
香草碎	2 茶匙

芝士汁

車打芝士片	4 片
牛油	25 克
砂糖	1 茶匙
雞粉	1/3 茶匙
麵粉	2 茶匙
清水	80 毫升

Liu's tips

- 花蝦灼水後必須吸乾水分，才淋上芝士汁，水分太多會令芝士汁不黏稠。

- 烘焗花蝦的時間不宜太久，容易令肉質過老。

做法

1. 預備芝士汁：慢火煮溶牛油，徐徐加入麵粉拌勻，注入清水煮熱，再下其他材料煮溶，盛起備用。

2. 煙肉切粒；洋葱橫切成圓片狀，放在焗盤上。

3. 大花蝦的蝦身中段位置去殼（留頭、留尾），在蝦身中央剗開，洗淨，瀝乾水分。蝦尾向下從蝦身缺口穿上，輕輕拉直成直線，用滾水稍灼定型。

4. 大花蝦用布吸乾水分，放在焗盤的洋葱片上，在蝦身淋上芝士汁，灑上煙肉碎及香草碎，放入已預熱 10 分鐘的焗爐，用 220℃ 焗約 5 至 6 分鐘至表面呈金黃色即可。

Deep-fried Fish in Sweet and Sour Sauce

糖醋菊花魚

Liu's tips

- 當魚肉縱橫切花時，剞至皮層深度即可，而且剞位愈深，魚肉炸後花紋愈清晰，呈菊花形狀，但小心切斷皮層。

- 撲粟粉時，留意魚肉每個位置需均勻地沾上粟粉，否則菊花魚欠鬆脆口感。

- 下油鑊前先將魚肉抖動一下，令多餘的粟粉落下，讓每段魚肉分開來，炸出來的形狀較明顯。

- 炸魚時油溫不能太低，否則粟粉很易脫落分離。

材料

鯇魚肉	600 克
糖醋汁	150 毫升
（參考 p.7）	

料頭

三色椒粒	80 克
蒜茸	半茶匙
葱粒	1 茶匙

醃料

鹽	1 茶匙
雞粉	半茶匙
麻油	2 茶匙
胡椒粉	半茶匙
雞蛋	1 個
粟粉	3 茶匙 （後下）

做法

1. 鯇魚肉洗淨,抹乾水分,先橫切後直切剞成花,深度約至皮層(不切斷),然後切件。

2. 魚肉用醃料拌勻醃約 15 分鐘,均勻地撲上粟粉。

3. 鑊內注入生油半鑊,燒熱至中油溫(約 150℃),放入魚肉炸熟至金黃色,盛起,瀝去油分,上碟。

4. 煮熱糖醋汁,加入料頭炒煮,下生粉水 2 茶匙勾茨煮熟,扒在魚件上即可。

糖醋菊花魚

A Palatable Taste · Desserts
滋味。甜點

一台焗爐、一個鑊，
裝備毋須太多，也可飯後嘗甜，
滋味滿足！

With just an oven or a wok, it's as simple as that,
one can taste the sweetness in the finale of a meal.
It's so beautiful and satisfying!

Deep-fried
Sweet Potato Dumplings

薯茸黃金角

材料

糯米粉	150 克
日本番薯	200 克
澄麵	80 克
滾水	80 毫升
砂糖	20 克
吉士粉	2 茶匙
生油	1 茶匙
清水	90 毫升

餡料

黃砂糖	3 湯匙
白芝麻	2 湯匙
炸花生碎	100 克
椰絲	2 湯匙
花生醬	2 湯匙

Liu's tips

- 鑊內之炸油需要蓋過黃金角，才可以令炸角有足夠的空間浮起，不易黏底燒焦。

- 炸油可先用中火燒熱，再調回小火炸黃金角，最後再用中火炸至皮脆。

做法

1. 餡料攪拌至均勻，備用。

2. 澄麵 80 克用滾水 80 毫升拌勻，搓成熟粉糰。

3. 番薯蒸熟，去皮，搓成薯茸。

4. 糯米粉加入清水拌勻，下吉士粉、薯茸、砂糖及熟澄麵糰，搓揉成粉糰，最後加入生油搓勻，即成皮料。

5. 將皮料分成每粒約 25 克的小粉糰，壓成圓形，包入 1 茶匙餡料，捏成三角形。

6. 鑊內下生油燒熱，待油溫約 120℃放入黃金角炸約 3 分鐘，至微微浮起後轉中火炸至表面金黃色及外皮香脆，盛起，趁熱享用。

薯茸黃金角

Glutinous Rice Balls Stuffed with Egg Custard

奶皇糯米糍

糯米糍皮料

糯米粉	150 克
澄麵	40 克
砂糖	120 克
椰奶	200 毫升
鮮奶	100 毫升
椰絲	200 克

奶皇餡料

鮮奶	250 毫升
椰奶	200 毫升
砂糖	90 克
魚膠粉	25 克
吉士粉	40 克
奶粉	30 克
牛油	40 克
清水	120 毫升
雞蛋	1 個（拂勻）

做法

1. 糯米糍皮料拌勻（椰絲除外），倒入蒸盆內，隔水蒸約半小時，待涼備用。

2. 將奶皇餡料的砂糖及魚膠粉混合，加入清水拌勻，倒入蒸盆內蒸溶（約 10 分鐘），取出，攪勻備用（見圖 1 及 2）。

3. 牛油蒸溶後，下所有奶皇餡料及魚膠粉溶液拌勻，過濾，倒入蒸盆內蒸約半小時，取出待涼，放於雪櫃冷藏半小時取出，分成小粒，方便包餡時使用。

4. 糯米皮搓成長條形，切成小粒，壓平後包入奶皇餡料，搓成小圓粒，滾上椰絲即可。

・奶皇餡料蒸熟後應拌勻，待涼後冷藏約半小時，使餡料不鬆散，易於包裹。

・將糯米皮搓成長條時如有黏手，可灑上椰絲再搓，避免出現黏手的情況。

奶皇糯米糍

Baked Sago Pudding

西米焗布甸

材料

西米	70 克
蓮蓉	160 克
砂糖	60 克
牛油	20 克
淡奶	40 毫升
鮮奶	80 毫升
雞蛋黃	2 個
吉士粉	30 克
清水	300 克

做法

1. 西米放入滾水內煮滾，熄火，加蓋浸焗至全粒透明，盛起，放入冰水或凍水內漂冷。

2. 每個焗盅內放入蓮蓉 20 克。

3. 煮熱清水，加入砂糖、牛油、淡奶、鮮奶及蛋黃煮熱，吉士粉用清水 40 毫升調溶勾獻攪勻，最後加入西米拌勻，放入焗盅內。

4. 排入已預熱 15 分鐘的焗爐內，用 180℃ 焗約 20 分鐘至表面呈金黃色即可。

- 西米不可用滾水煲煮，會令西米溶爛，不能形成透明粒狀，需要用滾水浸焗至發脹，浸凍水後使西米透明軟腍。

- 如使用玻璃焗盅，入爐前在焗盤上注入少許清水，再排上玻璃焗盅，可避免玻璃焗盅在高溫下有爆裂的情況。

西米焗布甸

Mini Walnut Cookies

小小合桃酥

材料

低筋麵粉	300 克
砂糖	100 克
固體油	150 克
合桃	60 克
雞蛋	1 個
雞蛋液	1 個
清水	40 毫升
食用梳打粉	2 克
泡打粉（發粉）	4 克

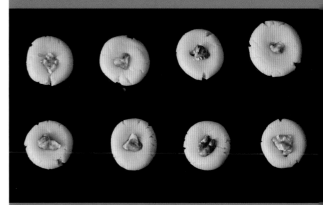

做法

1. 低筋麵粉用密篩隔勻，備用。

2. 砂糖、固體油、食用梳打粉及發粉拌勻至砂糖溶化，加入雞蛋 1 個及清水搓勻，慢慢加入麵粉搓勻至麵糰光滑，蓋上保鮮紙靜待20 分鐘。

3. 將麵糰搓成長條，切成小麵糰，搓圓後放於焗盤上。

4. 將小麵糰輕輕按壓，於中央按下小窩，放上合桃，掃勻蛋液。

5. 焗爐調至 180℃ 預熱 15 分鐘，將爐面火保持 180℃，底火轉至 100℃，放入合桃酥焗約 15 分鐘至呈少許金黃色，取出，塗上蛋液。焗爐的面火調低至 150℃，底火關掉，再焗 5 分鐘後，關掉所有火力，待 3 分鐘後取出享用。

Liu's tips

- 拌粉糰時，不宜用力搓擦，輕壓粉糰即可，否則容易使粉糰產生筋度，焗製後口感較硬、欠鬆脆。

- 若家中的焗爐沒有底面火溫度分別調校，建議用底面火160℃焗 15 分鐘，取出塗蛋液後轉爐面火 150℃，底火關掉，再烘焗 5 分鐘，熄火待 3 分鐘。

- 合桃酥排在焗盤上不要太密，要預留位置發酵脹大。

- 合桃酥焗製後宜完全涼透，使熱力揮發泡打粉之氣味才食用。

Lotus Seed and Mung Bean Sweet Soup

蓮子綠豆爽

材料

去殼綠豆	200 克
蓮子	120 克
馬蹄肉	120 克
薑	2 片
雞蛋	2 個（拂勻）
清水	6 杯
馬蹄粉	25 克

調味料

冰糖	160 克

做法

1. 蓮子及去殼綠豆分別用清水浸泡，蓮子浸軟、去掉芯；綠豆用清水滾約 15 分鐘，隔水備用；馬蹄肉洗淨，切薄片。

2. 煲內放入清水，下薑片及蓮子煲約 45 分鐘至蓮子軟腍，即可加入綠豆及馬蹄肉煲滾，再下冰糖調味。

3. 用適量清水調溶馬蹄粉，倒入糖水內攪勻勾茨，煮滾後熄火，輕輕攪動糖水，並逐少加入雞蛋液成蛋花，即可享用。

Liu's tips

- 煲蓮子時不可加入糖，否則蓮子難以煲腍，需待蓮子完全腍身後才可加入冰糖。

- 蛋液宜熄火後才加入並攪成蛋花，不宜用猛火煮熟，否則蛋液過火不嫩滑。

蓮子綠豆爽

Water Chestnut Cake with Sugarcane and
Couchgrass Root Flavour

竹蔗茅根馬蹄糕

Liu's tips

.   .

若煮滾的竹蔗茅根汁溫度太熱，與粉料拌勻時，會令粉料過熟變稠，最後倒入糕盆後令表面不平，影響外觀。

蒸熟的馬蹄糕可凍食或熱食，喜熱食的話建議切件煎香，或於雪櫃取出後，切件翻蒸亦可。

材料

馬蹄肉	150 克
馬蹄粉	120 克
吉士粉	3 克
椰糖	100 克
清水	200 毫升
竹蔗茅根汁	500 毫升
（超市支裝飲料）	

做法

1. 馬蹄肉洗淨，切薄片備用。

2. 馬蹄粉、吉士粉及清水放在盆內調溶。

3. 竹蔗茅根汁與椰糖煮滾，加入馬蹄片續煮。

4. 糕盆內掃上薄薄的油；將微暖的竹蔗茅根汁倒入已調勻之粉料拌至漿狀，倒入糕盆內蒸約 30 分鐘，待涼後切件享用。

竹蔗茅根馬蹄糕

海底椰栗子煲豬腱

Coco-de-Mer, Chestnut and Pork Shin Soup

Ingredients
30 g dried coco-de-mer
200 g chestnuts
300 g spareribs
400 g pork shin
3 big slices ginger
4 candied dates
8 cups water

Seasoning
1/2 tsp salt

Method
1. Rinse the dried coco-de-mer and soak in water for about half an hour.
2. Remove the shell and skin of the chestnuts.
3. Rinse the pork shin and spareribs, blanch in boiling water and set aside.
4. Bring water to the boil, put in all the ingredients and cook over high heat for about half an hour. Turn to medium heat and cook for about 2 hours. Season with salt and serve.

Liu's tips
· Soak the dried coco-de-mer in water before cooking. It is easier for the softened coco-de-mer to release its flavour.
· Put the shelled chestnuts into a microwave oven and heat over medium for 1 minute. The skin can be easily removed.

Cookware:
Large stainless steel stockpot

Double-steamed Black-skinned Chicken Soup with Agrocybe Aegerila Mushroom and Dried Scallop

茶樹菇瑤柱燉竹絲雞

Ingredients
1/2 black-skinned chicken
60 g dried Agrocybe Aegerila mushrooms
60 g dried scallops
20 g dried longan
80 g diced lean pork
2 slices ginger
4 cups water

Seasoning
1/2 tsp salt

Method

1. Soak the dried Agrocybe Aegerila mushrooms in water for about half an hour, rinse and remove grit. Set aside.
2. Cover the dried scallops with water and soak for about 45 minutes to soften.
3. Rinse the black-skinned chicken and lean pork, blanch in boiling water, put into a container for double-steaming. Add the Agrocybe Aegerila mushrooms, dried longan, ginger and dried scallops (with the soaking water), and then fill with water. Put a lid on, seal the lid and double-steam for about 3 hours. Season with salt. Serve.

Liu's tips

· Before cooking, soak the dried scallops in water until they swell and become softer. This will help infuse their fresh flavour into the soup.
· You may use fresh Agrocybe Aegerila mushrooms and cook them right after they are bought. They will easily rot if they are kept for too long.

Cookware:
Large stainless steel stockpot or wok

Double-steamed Monkey-head Mushroom Soup with Jinhua Ham and Winter Melon

Ingredients
50 g monkey-head mushrooms
400 g winter melon
50 g Jinhua ham
80 g lean pork
20 g dried wolfberries
8 small slices ginger
1.5 cups chicken stock
1.5 cups water

Seasoning
1/2 tsp salt

Method
1. Soak the monkey-head mushrooms in water for about 1 hour, wash away dirt and impurities from the surface and squeeze water out. Rinse the dried wolfberries and set aside.
2. Cut the winter melon with skin on into thick slices. Dice the Jinhua ham and lean pork, blanch in boiling water and set aside.
3. Mix the chicken stock with water, bring to the boil and set aside.
4. Put all the ingredients into a container for double-steaming, pour in the boiled chicken soup, put a lid on and seal the lid. Double-steam for about 2 hours and 30 minutes, season with salt. Serve.

Liu's tips
· The monkey-head mushrooms must be squeezed dry after they are soaked and rinsed to remove their odd smell; otherwise, the flavour of soup will be affected.
· If you want a beautiful presentation of the soup in the banquet, you may thinly slice the Jinhua ham, cut into the middle of the winter melon slice (do not cut off) and then place the ham inside the winter melon, like a "ham and winter melon sandwich". Arrange them into the container and then double-steam the soup.

Cookware:
Large stainless steel stockpot or wok

Beef Thick Soup with Egg and Spinach

蛋花菠菜牛肉羹

Ingredients
300 g minced beef
300 g spinach
6 fresh black mushrooms
1 stalk coriander
1 sprig spring onion
1 egg (whisked)
2 cups chicken stock
2 cups water

Marinade
1/2 tsp salt
2 tsp caltrop starch
6 tsp water
1/4 tsp ground white pepper

Seasoning
1 tsp salt
1/4 tsp ground white pepper
1/2 tsp sesame oil

Thickening glaze
2 tbsp caltrop starch
6 tbsp water

Method
1. Mix the minced beef with the marinade, rest for half an hour and blanch in boiling water. Set aside.
2. Rinse the fresh black mushrooms, finely dice and set aside.
3. Rinse the spinach, finely chop and blanch in boiling water until soft. Rinse in cold water and squeeze water out.
4. Finely dice the coriander and spring onion. Set aside.
5. Pour the chicken stock and water into a stockpot, bring to the boil, add the beef, fresh black mushrooms and spinach, and then heat up. Stir in the seasoning and thickening glaze, gently mix in the egg wash, finally sprinkle with the coriander and spring onion. Serve.

Liu's tips
· To remove the grassy smell of spinach and keep it green, slightly blanch the spinach in boiling water and rinse it in cold water.
· The secret of keeping the minced beef smooth is to marinate it with caltrop starch.

Cookware:
Large stainless steel stockpot

145

雪梨排骨生魚湯

Snakehead Soup with Pears and Spareribs

Ingredients
3 pears
1 snakehead (about 450 g)
300 g spareribs
3 slices ginger
8 red dates (cored)
8 cups water

Seasoning
1/2 tsp salt

Method
1. Peel the pears, cut into small pieces, remove the cores, and then cover with water and set aside.
2. Blanch and rinse the spareribs.
3. Bring water to the boil, put in the pears, spareribs and red dates, and then bring to the boil.
4. Rinse the snakehead and wipe dry. Heat a wok, add about 2 tbsp of oil, put in the ginger and snakehead, fry until both sides of the fish are light brown. Put them into a filter bag, add into the boiling soup and cook for about half an hour. Turn to medium heat and cook for 1 hour and 30 minutes.
5. Skim the oil from the surface of the soup, transfer the soup ingredients to a plate, season the soup with salt. Serve.

Liu's tips
The pears must be cored before cooking because the core of fruit when heated will give a sour taste, that will affect the flavour of soup.

Cookware:
Large stainless steel stockpot

Double-steamed Pigeon Soup with Conch and Dried Scallop

螺頭元貝燉乳鴿

Ingredients
2 pigeons
300 g frozen conch meat
60 g dried scallops
80 g diced lean pork
30 g sweet and bitter almonds
4 large dates
2 slices ginger
4 cups water

Seasoning
1/2 tsp salt

Method
1. Cover the dried scallops with water and soak for about 45 minutes to soften.
2. Cut open the frozen conch meat, remove the intestine, rub with 2 tsp of caltrop starch and 1 tsp of salt, wash and cut into coarse strips.
3. Rinse the pigeons and lean pork, blanch with the conch meat in boiling water. Set aside.
4. Put all the ingredients into a container for double-steaming, fill with water, put a lid on and then double-steam for about 3 hours. Season with salt and serve.

Liu's tips
Rub the conch meat with caltrop starch and salt to remove dirt and the smell of frozen seafood, so that the fresh and sweet flavour of the soup can be retained.

Cookware:
Large stainless steel stockpot or wok

147

滑豆腐魚雲羹

Tofu and Fish Head Thick Soup

Ingredients
1 bighead carp's head
(about 800 g)
1 soft tofu
160 g roasted lean pork
8 Chinese flowering
cabbage stems
80 g carrot
1 egg (whisked)
2 cups chicken stock
2 cups water
2 slices ginger
1 stalk coriander
1 sprig spring onion

Seasoning
1 tsp salt
1/4 tsp ground white pepper
1/2 tsp sesame oil

Thickening glaze
2 tbsp caltrop starch
6 tbsp water

Method
1. Cut open the bighead carp's head and then rinse. Arrange the ginger on the head, steam until cooked, then remove the bone and keep the meat. Loosen the meat by hand. Set aside.
2. Thinly slice the Chinese flowering cabbage stems and carrot. Set aside.
3. Cut the roasted lean pork into thin slices. Finely dice the tofu.
4. Finely dice the coriander and spring onion. Set aside.
5. Heat the chicken stock with water, put in all the ingredients, bring to the boil and mix in the seasoning. Stir in the thickening glaze, heat up, stir in the egg wash to decorate, sprinkle with the coriander and springs onion. Serve.

Liu's tips
· You should be careful of removing the bone of the steamed fish head. All the tiny bones must be removed to avoid choking on them while eating.
· To make the soup silkier, all the ingredients need to be finely cut.
· You may use minced pork to replace the roasted lean pork to suit your taste.

Cookware:
Large stainless steel stockpot

Coconut, Corn and Chicken Soup

椰子粟米煲雞湯

Method

1. Soak the white fungus and almonds in water for about half an hour. Cut away the hard stalks in the middle of the white fungus with scissors.
2. Skin the carrot and cut into small pieces. Remove the husk from the corn, rinse and chop into pieces.
3. Rinse the chicken and spareribs, blanch in boiling water and set aside.
4. Break open the coconut, pour out the coconut water and keep it. Cut the coconut flesh into small pieces and rinse.
5. Put all the ingredients (except the white fungus) and coconut water into a stockpot, cook over high heat for about half an hour. Add the white fungus, turn to low-medium heat and cook for about 1 hour and 30 minutes. Turn off the heat. Transfer the soup ingredients to a dish, skim the oil from the surface of the soup and season with salt. Serve.

Ingredients
1 shelled coconut
2 fresh corn on the cob
1 white fungus
1 carrot
1/2 chicken
300 g spareribs
80 g sweet and bitter almonds
10 cups water

Seasoning
1/2 tsp salt

Liu's tips
It is easier to cook the white fungus tender after it is softened and swollen. White fungus is not suitable to be cooked over high heat. Add it to the soup only after the heat is adjusted to low-medium.

Cookware:
Large stainless steel stockpot

149

家鄉釀涼瓜 Stuffed Bitter Cucumber

Ingredients
2 bitter cucumbers

Pork filling
400 g minced pork
40 g dried shrimps
1 egg

Marinade
1/2 tsp salt
1/2 tsp chicken bouillon powder
2 tbsp caltrop starch

Spices
1 tbsp fermented black bean sauce
(refer to p.8)
1 tsp finely chopped garlic
1 tsp finely chopped ginger
20 g diced spring onion

Seasoning
1/2 tsp salt
1 tbsp oyster sauce
1 tsp sugar
1 tbsp light soy sauce
1 tsp dark soy sauce
1 cup water

Cookware:
Non-stick wok

Method
1. Cut across the bitter cucumbers into ringlike chunks of 2 cm thick, scrape off the seeds with white pith, blanch in hot water for about 3 minutes. Dish up, cool in cold water, sop up the moisture and set aside.
2. Soak the dried shrimps in water until soft, finely chop, mix with the other ingredients of the pork filling, and stir until gluey. Spread caltrop starch on the inner rings of the bitter cucumbers and stuff with the pork filling.
3. Heat a wok, add 2 tbsp of oil and fry the stuffed bitter cucumbers over low heat until both sides are golden. Dish up.
4. Sauté the spices in the wok until fragrant, add the seasoning and heat up. Put in the bitter cucumbers and simmer until cooked. Finally thicken the sauce with 2 tbsp of caltrop starch solution. Serve.

Liu's tips
· Before stuffing, spread a little dry caltrop starch on the inner rings of the bitter cucumbers to prevent the filling from falling off during cooking.
· Fry the stuffed bitter cucumbers until the pork filling is half-cooked. Do not cook them with the seasoning for too long. This is to avoid overcooking the bitter cucumbers which will easily fall apart and look withered and yellow.

Crisp Mud Carp Balls

脆皮鯪魚球

Ingredients
20 tofu puffs
300 g minced mud carp
1/3 dried tangerine peel
30 g dried shrimps
1 sprig spring onion
2 stalks coriander

Method
1. Soak the dried shrimps and tangerine peel in water until soft. Rinse the spring onion and coriander. Finely dice all the above ingredients, add the minced mud carp, and stir in one direction until gluey.
2. Cut open the middle of the tofu puffs with scissors (do not cut off), turn the inside out, stuff with the mud carp filling, and tightly press to seal the puffs.
3. Heat about 3 cups of oil in a wok over medium heat, deep-fry the stuffed tofu puffs until cooked and crisp (around 5 to 6 minutes), drain the oil. Serve hot.

Liu's tips
· You can buy seasoned minced mud carp in the market. It has enough salty taste and so no additional salt is needed. Just mix it with the other ingredients.
· Try giving a good control of the temperature of oil for deep-frying the tofu puffs. They will end up with a tough surface if the heat is too high.

Cookware:
Wok or saucepot

Sichuan Spicy Boiled Grass Carp Belly

水煮鯇魚腩

Ingredients

500 g grass carp belly meat
160 g soy bean sprouts
100 g Chinese celery
1 small cucumber
80 g dried red chilli
3 tbsp Sichuan peppercorns
8 tbsp chilli oil
4 slices ginger
8 garlic cloves
3 stalks coriander

Marinade

1/2 tsp salt
1 egg white
2 tsp caltrop starch
1 tsp white vinegar
1 tbsp Sichuan pepper oil

Seasoning

2 tbsp Sichuan-styled chilli sauce
2 tbsp light soy sauce
1 tbsp dark soy sauce
1.5 tsp salt
1 tsp chicken bouillon powder
1 tbsp sugar
1 tbsp Sichuan peppercorns
1 tbsp Sichuan pepper oil
2 bird's eye chillies
6 cups water

Method

1. Rinse the grass carp belly meat, cut into slices, mix with the marinade and set aside.
2. Slice the cucumber. Section the Chinese celery and rinse.
3. Put Sichuan pepper oil in a wok, sauté the ginger, garlic and Sichuan peppercorns until fragrant, add the soy bean sprouts and stir-fry until aromatic. Put in the cucumber and seasoning, cook for about 3 minutes. Add the fish slices and Chinese celery, boil until cooked and then pour into a clay pot, set aside.
4. Heat the chilli oil in the wok, sauté the dried chilli and Sichuan peppercorns until fragrant, pour into the pot and then sprinkle with the coriander. Serve.

Liu's tips

· The fish marinated with white vinegar will be a bit firmer and will not easily fall apart during cooking.
· Turn off the heat once the soup boils after adding the fish, so that the fish will not be overcooked.

Cookware:
Wok

Steamed Egg with Shrimps

鮮蝦蒸水蛋

Ingredients
3 eggs
6 fresh shrimps
1 tsp diced spring onion

Seasoning
1/2 tsp salt
270 ml warm water

Method
1. Remove the heads and shell of the shrimps, keep the tails, slit open the shrimp back, pick out the intestine, rinse and then blanch in boiling water. Set aside.
2. Whisk the eggs, mix in the seasoning, and sieve the egg wash with a fine meshed sieve. Pour into a deep dish and steam over high heat for about 2 minutes.
3. Remove the lid, add the shrimps, turn to low heat and steam for about 4 to 5 minutes. Take out, pour in some cooked oil and light soy sauce, sprinkle the diced spring onions on top. Serve.

Liu's tips
· Prepare a shallow dish of egg wash and do not steam it all the way over high heat. First steam over high heat, then low heat.
· You can remove the cover during the steaming process to observe the doneness of the egg. It can also reduce the heat inside the wok to make the egg smooth.
· Keep in mind that the egg wash must be mixed with warm water, and the proportion of egg wash to warm water is 1 : 1.5. Do not mix it with cold water.

Cookware:
Wok

153

菜粒冬菇臘味飯

Preserved Meat Rice with Vegetable and Dried Black Mushrooms

Ingredients
2 preserved pork sausages
1 preserved liver sausage
1/3 piece preserved pork
120 g Chinese kale
6 dried black mushrooms
300 g white rice

Mixed soy sauce
3 tbsp light soy sauce
2 tsp dark soy sauce
1 tsp sugar
1 tsp oil
1/2 tsp sesame oil
3 tbsp boiled water
ground white pepper

Method
1. Blanch the preserved pork sausages, liver sausage and pork in boiling water, cut into dices.
2. Soak the dried black mushrooms in water until soft, remove the stalks and finely dice. Rinse the Chinese kale, cut into dices and set aside.
3. Heat the mixed soy sauce and set aside.
4. Rinse the white rice, put into a rice cooker, add 300 ml of water and start cooking. When the water starts to dry after around 15 minutes, lay the preserved meat and black mushrooms on the rice, put the lid on and cook for about 5 minutes. Finally add the Chinese kale and cook until the rice is done.
5. Season with the mixed soy sauce and give a good stir before serving.

Liu's tips
· Do not put in the diced vegetable too early, or it will be overcooked and turn brown. The presentation is not good.
· You may add 1 stalk of coriander to cook with the mixed soy sauce to give it a hint of coriander aroma.

Cookware:
Rice cooker

Chicken and Tofu in Casserole

雞粒滑豆腐煲

Ingredients
2 cloth-wrapped tofu
200 g boned chicken thigh
4 dried black mushrooms
1 small piece Mei Xiang
salted fish
1/2 cup chicken stock

Spices
finely chopped garlic
sliced ginger
diced spring onion

Marinade
1/2 tsp salt
2 tsp caltrop starch
2 tbsp water
1/2 tsp sesame oil

Seasoning
1/3 tsp salt
2 tsp oyster sauce
1/2 tsp sugar
1 tsp dark soy sauce
1 tsp Chu Hou sauce
2 tsp caltrop starch
1 tbsp water

Method
1. Skin the boned chicken thigh, rinse and dice, mix with the marinade.
2. Soak the dried black mushrooms in water until soft, remove the stalks, rinse and blanch in boiling water. Rinse in cold water and finely dice.
3. Fry the salted fish over low heat until fragrant, and then dice, set aside.
4. Dice the tofu, cover with hot water, add 1/2 tsp of salt and mix well.
5. Heat 1 tbsp of oil in a wok, stir-fry the chicken until cooked, put in the spices and sauté until aromatic. Add the chicken stock, black mushrooms and salted fish, and give a good stir-fry.
6. Drain the tofu, put into the wok and simmer for 1 minute. Add the seasoning, stir to thicken the sauce, bring to the boil and transfer to a casserole. Serve.

Liu's tips
· The tofu soaked in hot salted water will be firmer on the outside and not easily fall apart during cooking.
· Fry the salted fish first to make it more fragrant.

Cookware:
Wok

Braised Mandarin Fish with Radish

蘿蔔燜桂花魚

Ingredients
1 mandarin fish (about 500 g)
250 g white radish
8 dried black mushrooms
10 garlic cloves
sliced ginger
shredded spring onion
shredded red chilli

Seasoning
1 tsp salt
1/2 tsp sugar
2 tbsp oyster sauce
1 tsp sesame oil
1/2 tsp ground white pepper
1 tbsp Chu Hou sauce
1 tbsp cooking wine
2 cups water

Method
1. Soak the dried black mushrooms in water until soft (about 2 hours), remove the stalks. Rinse well, blanch in boiling water and cut in half, set aside.
2. Rinse the mandarin fish, wipe dry, score the body in a crisscross pattern.
3. Skin the white radish, cut into thick slices and blanch in boiling water for about 5 minutes. Set aside.
4. Thinly spread caltrop starch on the mandarin fish, deep-fry over medium heat until fully cooked and dish up. Put in the garlic and deep-fry until golden, drain well.
5. Sauté the ginger in the wok until fragrant, add the seasoning and heat up. Put in the white radish, black mushrooms and garlic, cook over medium heat for about 20 minutes.
6. Turn to low heat, add the mandarin fish and simmer for about 6 minutes. Put on a dish and arrange the white radish and mushrooms on the side. Thicken the sauce with caltrop starch solution, pour onto the fish, sprinkle the shredded spring onion and red chilli on top. Serve.

Liu's tips
· The fish will easily cook through by scoring the body crisscross before deep-frying.
· It will keep the braised garlic in shape by deep-frying it beforehand.

Cookware:
Wok

Shrimp Stuffed Squids with Pepper Salt

椒鹽百花釀鮮魷

Ingredients
200 g frozen shrimp meat
2 fresh squids (about 600 g)
6 shelled water chestnuts
1/2 tsp pepper salt

Spices
finely chopped garlic
diced red chilli

Seasoning
1/3 tsp salt
1/3 tsp chicken bouillon powder
3 tsp caltrop starch
1/2 tsp sesame oil
ground white pepper

Method

1. Finely chop the shelled water chestnuts. Defrost the shrimp meat, rinse and drain well, sop up the moisture with a dry cloth. Put on a cutting board, bash with the blade of a knife, then chop with the spine of the knife into puree.
2. Put the shrimp puree into a bowl, stir by hand until a bit sticky, add the seasoning, stir until gluey, throw into the bowl repeatedly until spongy. Chill the minced shrimp in a fridge.
3. Cut open the body of the squids with scissors. Remove the guts, skin and tentacles, wash thoroughly. Score the body surface crisscross to make a pattern, then cut into small pieces in a rectangular shape (2 cm x 4 cm). Blanch in boiling water, sop up the moisture, spread a thin layer of caltrop starch onto the plain side, stuff with the minced shrimp.
4. Heat about 2 cups of oil, deep-fry the stuffed squids over medium heat until cooked and golden. Set aside.
5. Dish up the oil, put in the spices and sauté until fragrant, add the stuffed squids and give a good stir-fry. Finally sprinkle with the pepper salt and mix well. Serve.

Liu's tips

· To make tasty minced shrimp, the key is to sop up the moisture of the shrimp meat. Also, it must be stirred in one direction to make it sticky and elastic.
· The purpose of blanching the squids is to give shape so that the filling will not fall off easily. It has the same effect by spreading caltrop starch onto the squids before stuffing.
· Just deep-fry the stuffed squids over medium heat for about 3 minutes. The squids will shrink a lot by long deep-frying, making the shrimp filling look over-sized. The presentation will be less beautiful.

Cookware:
Wok

Fried Meat Patties with Salted Fish and Lotus Root

鹹魚煎藕餅

Ingredients
250 g minced pork
80 g minced mud carp
60 g Mei Xiang salted fish
150 g lotus root
2 tsp diced spring onion
2 tsp diced ginger

Seasoning
1/3 tsp salt
1 tbsp oyster sauce
1 tsp sugar
1 tsp sesame oil
ground white pepper
2 tbsp caltrop starch
1/2 egg wash

Method
1. Peel the lotus root, finely dice, rinse and drain.
2. Slowly fry the salted fish until fragrant, chop up, pour into a big bowl, add the minced pork, minced mud carp and seasoning, and stir until gluey.
3. Finally add the diced lotus root, spring onion and ginger, mix well. Squeeze the pork mixture to make little meat balls, put onto a greased plate, then slightly flatten to make meat patties.
4. Heat a wok, add 3 tbsp of oil, slowly fry the meat patties until both sides are cooked through and golden. Serve.

Liu's tips
· The salted fish must be fried and chopped before mixed with the other ingredients; otherwise, its savoury smell can hardly come out.
· Just buy the seasoned minced mud carp in the market to make this dish.

Cookware:
Non-stick wok

Baked Mackerel Steaks in Teriyaki Sauce

燒汁煎焗鮫魚扒

Ingredients
2 mackerel steaks
1/2 shredded onion
3 tbsp teriyaki sauce
salad dressing

Marinade A
3 tbsp light soy sauce
2 tbsp dark soy sauce
1 tsp salt
2 tsp chicken bouillon powder
2 tsp sugar
1 tbsp Hua Diao wine
1/2 tsp ground white pepper
4 bay leaves
1 cup water

Marinade B
150 g Chinese celery
150 g carrot
4 shallots
1/2 onion
10 g ginger
3 stalks coriander
1 cup water
2 lime (juice squeezed)

Method
1. Simmer the marinade A until dissolved, leave to cool down.
2. Blend the marinade B in a blender, sieve the sauce, mix the marinade A with the sauce of B, put in the rinsed mackerel steaks and soak for about 4 hours.
3. Wipe the mackerel steaks dry, heat a wok, add some oil, fry the mackerel steaks over medium heat until both sides are fragrant, set aside.
4. Preheat an oven to 240°C, line a baking tray with aluminum foil, lay with the shredded onion, brush the mackerel steaks with the teriyaki sauce, and put on top of the onion. Bake for about 5 minutes, take out, brush again with the teriyaki sauce, turn over and bake for 5 minutes, or until cooked through. Serve with the salad dressing.

Liu's tips
· Do not cut the mackerel steaks too thin (the desirable thickness is around 2 cm), or it will shrink and easily fall apart after heated.
· If you want a crisp surface, spread a pinch of flour on both sides of the mackerel steak and then fry until golden before baking.

Cookware:
Wok and oven

Deep-fried Spareribs and Pumpkin in Apricot Sauce

金瓜黃梅骨

Ingredients
300 g spareribs
300 g Japanese pumpkin
50 g corn chips

Marinade
1/2 tsp salt
1/2 tsp chicken bouillon powder
2 tsp caltrop starch
2 tbsp water
1 tsp custard powder
2 tbsp flour (added last)

Ingredients of apricot sauce
6 tbsp apricot jam
2 Thai limes
1 tbsp sugar

Method
1. Chop the spareribs into long sections, rinse and drain, mix with the marinade and leave for half an hour.
2. Rinse the Japanese pumpkin and cut into a rib-like shape with skin on. Grate the green part of the lime zest for later use, squeeze out about 1 tbsp of lime juice.
3. Slowly heat the apricot jam, lime juice and sugar. Add the lime zest, and stir to make the apricot sauce.
4. Heat about 2 cups of oil in a wok, mix the spareribs with some flour, put into the oil and deep-fry over medium heat until fully cooked, drain. Put in the pumpkin and deep-fry until crisp.
5. Heat the apricot sauce, mix in the spareribs and pumpkin, dish up and sprinkle the slightly crumbled corn chips on top. Serve.

Liu's tips
The skin of Japanese pumpkin is thin so you can leave the skin on. However, if it is a Chinese pumpkin, the skin should be removed to keep the soft texture of the flesh.

Cookware:
Wok

Steamed Meat Patty with Salted Egg and Water Chestnut

Ingredients
300 g pork collar-butt
1 salted egg
6 skinned water chestnuts
4 dried black mushrooms

Seasoning
1/3 tsp salt
1 tbsp oyster sauce
2 tsp caltrop starch
2 tbsp water

Method
1. Rinse the pork collar-butt, cut into small pieces and finely chop. Bash the water chestnuts and chop up. Soak the dried black mushrooms in water until soft and remove the stalks, finely dice.
2. Separate the salted egg white from the yolk. Flatten the yolk with a knife to a round shape. Reserve the egg white for seasoning.
3. Put the pork into a big bowl, add the seasoning and salted egg white, stir on clockwise until the meat is gluey.
4. Throw the meat into the bowl 5 to 6 times, then put on a plate. Flatten the meat, make a small hole in the middle and put in the yolk.
5. Steam the meat in a wok for about 10 minutes, take out, sprinkle diced spring onion on top and serve.

Liu's tips
· You may buy minced pork prepared by a food processor, but the meat texture is relatively loose and not so resilient.
· Do not stir and throw the seasoned pork for too many times, or you will end up with a shrunk and firm pork patty with poor mouthfeel.
· Flatten the meat into a thin layer before steaming it. If too thick, it can hardly cook through and will not have a nice meat texture.
· It is more manageable by bashing the water chestnuts in a zip lock bag.
· Spread egg white on the knife before pressing the yolk to prevent it from sticking to the knife.

Cookware:
Wok

Baked Egg with Mussels in Earthen Bowl

Ingredients

12 frozen shelled mussels
50 g minced pork
30 g diced fat pork
1/3 dried tangerine peel
1 tsp finely chopped ginger
1 tsp finely chopped garlic
1 tsp diced spring onion
1 egg wash
1/4 piece deep-fried dough stick

Seasoning

3 eggs
1/2 tsp salt
1/3 tsp chicken bouillon powder
water (its proportion to whisked egg is 1 : 1)
1/2 tsp sesame oil
1/3 tsp ground white pepper

Method

1. Rinse the shelled mussels, blanch in boiling water, coarsely dice and put into an earthen bowl.
2. Soak the dried tangerine peel in water until soft, scrape off the pith and chop up. Slice the deep-fried dough stick.
3. Put the minced pork and diced fat pork into a wok, add the ginger, garlic and dried tangerine peel, stir-fry until fragrant, then put into the earthen bowl.
4. Mix the seasoning, add the diced spring onion, pour into the earthen bowl, arrange the deep-fried dough stick on top, steam for about 12 minutes, or until cooked. When it cools, sop up the moisture on the surface with a dry cloth or kitchen paper.
5. Preheat an oven to 240℃, brush the egg wash on the steamed mussel mixture, put into the oven and bake for about 8 minutes. Take out, brush with the egg wash again and bake for about 5 minutes, or until golden. Serve.

Liu's tips

· To save the time, you can steam the bowl of mussel mixture in advance and put it into a fridge when it cools. When you are ready to cook, take it out and heat up by steaming, and bake it after brushed with the egg wash.
· If a small earthen bowl is used, fill the bowl with 80% full of the mixed ingredients. If it is too full, there will not be enough space for you to brush the egg wash and the egg mixture will swell and spill after baked.

Cookware:
Wok and oven

梅菜扣肉

Steamed Pork Belly with Preserved Flowering Cabbage

Ingredients
500 g pork belly
300 g preserved flowering cabbage

Spices
1 tsp finely chopped garlic
1 tsp finely chopped shallot

Seasoning
2 tbsp Chu Hou sauce
1 tbsp red fermented beancurd
2 tbsp oyster sauce
1/2 tsp salt
1 tbsp sugar
1 tbsp light soy sauce
1 tsp dark soy sauce
2 tsp rose wine
1 cup water

Method

1. Tear the preserved flowering cabbage into small stalks, soak in water for about 1 hour to soften and reduce the salty taste. Wash away the grit, squeeze water out and cut into dices.
2. Blanch the pork belly in hot water over low heat for about 45 minutes, dish up, wash grease away, and spread dark soy sauce on the skin to give colour. Heat a wok, add a little oil, fry the pork belly skin until brown and set aside.
3. Heat the wok and stir-fry the preserved flowering cabbage without adding oil until fragrant. Put in about 2 tbsp of sugar, stir-fry until fragrant and light brown, add 2 tbsp of oil, mix up and set aside.
4. Cut the pork belly into thick slices. Heat some oil in the wok, sauté the spices, add the pork belly and seasoning and then sauté until fragrant. Arrange the pork belly on a big bowl, lay the preserved cabbage on top, and pour in the sauce.
5. Steam the bowl of pork belly for about 2 hours, take out, pour out the sauce, thicken the sauce with caltrop starch solution and pour over the pork belly. Serve.

Liu's tips

· Spread out the leaves of the softened preserved flowering cabbage and wash away the grit embedded in the leaves.
· Do not blanch the pork belly over high heat to avoid it being too tender. If the pork belly is too soft, the loose skin layer will easily fall apart while cutting it into pieces.

Cookware:
Wok

芝麻雞中翼 Baked Chicken Wings with Sesame Seeds

Ingredients
10 chicken mid-joint wings
80 g white sesame seeds
1 egg white
2 tsp cornstarch

Marinade
1 tbsp light soy sauce
2 tsp dark soy sauce
1 tsp sesame oil

Chinese marinade
6 star anise
1 tsp Sichuan peppercorns
1 slice cinnamon
10 slices sand ginger
10 bay leaves
4 slices ginger
2 sprigs spring onion
5 cups water

Seasoning
3 tsp salt
2 tsp chicken bouillon powder
3 tsp sugar
2 tbsp Hua Diao wine

Cookware:
Saucepot and oven

Method
1. Put the Chinese marinade into a small saucepot, simmer for about 1 hour, add the seasoning and cook until dissolved.
2. Defrost the chicken wings, mix with the marinade and leave for about 1 hour. Put into the hot Chinese marinade, turn off the heat, leave with a lid on for about 15 minutes, take out, and put onto an oven rack to drain.
3. Mix the egg white with the cornstarch, spread onto the chicken skin, sprinkle with the white sesame seeds and air-dry for about 2 hours.
4. Preheat an oven to 240℃, bake the chicken wings for about 10 minutes, turn over and bake again for 10 minutes, or until golden. Serve.

Liu's tips
· It is not necessary to turn on the heat when soaking the chicken wings in the Chinese marinade; otherwise, they will be overcooked and become too tender.
· You can save the time by drying the chicken wings with an electric fan at low wind speed.

Salted Fish, Abalone and Chicken Rice

Ingredients

6 fresh abalones
1 boned chicken thigh
1 small piece salted
 fourfinger threadfin (about 40 g)
4 dried black mushrooms
20 g shredded ginger
20 g diced spring onion
300 g white rice

Marinade

1/2 tsp salt
1 tbsp oyster sauce
3 tsp caltrop starch
3 tbsp water

Mixed soy sauce

3 tbsp light soy sauce
2 tsp dark soy sauce
1 tsp sugar
1 tsp oil
1/2 tsp sesame oil
3 tbsp boiled water
ground white pepper

Method

1. Remove the shells and guts of the abalones, rinse and drain, then coarsely dice.
2. Skin the chicken thigh, cut into dices, mix with the marinade, leave for about 15 minutes, and mix with the diced abalone.
3. Soak the dried black mushrooms in water until soft, remove the stalks and cut into dices. Heat the mixed soy sauce and set aside.
4. Rinse the white rice, put into a rice cooker, add 300 ml of water and start cooking. When the water starts to dry after around 15 minutes, lay the diced chicken, abalone and black mushroom on the rice, then put the salted fish on top and sprinkle with the shredded ginger. Put the lid on and cook until the rice is done. Leave with the lid on for around 5 minutes.
5. Before serving, sprinkle the diced spring onion on top, season with the mixed soy sauce and give a good stir. Serve.

Liu's tips

Mix the gutted abalones with caltrop starch and put them in a colander. Keep stirring and rubbing the abalones to remove dirt on the surface, and wash them thoroughly.

Cookware:
Rice cooker

西汁烤肋排

Roasted Spareribs in Western Style Sweet and Sour Sauce

Ingredients
600 g spareribs
6 tbsp Western style sweet
and sour sauce (refer to p.10)
1 tbsp mixed herbs

Marinade
1 tsp salt
1/2 tsp chicken bouillon powder
1 tsp sugar
3 tsp caltrop starch
1 egg yolk

Method
1. Chop the spareribs into pieces of 6 cm long. Rinse and mix with the marinade, leave for about 1 hour. Add 3 tbsp of the western style sweet and sour sauce and mix well.
2. Preheat an oven to 200℃, lay the spareribs on a baking tray lined with aluminum foil, bake for about 10 minutes, turn over and bake again for 10 minutes.
3. Brush the spareribs with the sweet and sour sauce again, sprinkle the mixed herbs over. Adjust the heat to 240℃, and bake each side of the spareribs for 5 minutes. Serve with the sweet and sour sauce.

Liu's tips
The Western style sweet and sour sauce can be prepared in advance and stored in a fridge. This sauce can be applied to dishes like Chinese Style Fried Beef Tenderloin, Peking Pork or Deep-fried Pork Chop. You might as well prepare some at home for use anytime.

Cookware:
Oven

Deep-fried Chicken and Lotus Root with Red Fermented Beancurd

Ingredients
1/2 chilled chicken
200 g lotus root
1 tsp Hua Diao wine

Spices
1 tsp finely chopped garlic
1 tsp diced onion
1 tsp diced spring onion
1 tsp diced red chilli

Marinade
2 large red fermented beancurd
1/2 tsp chicken bouillon powder
1 tbsp oyster sauce
1 tbsp sugar
1 tbsp finely chopped garlic
1 tbsp diced ginger
2 tsp caltrop starch
3 tbsp low-gluten flour (added last)

Method
1. Peel and rinse the lotus root, thinly slice. Take half cube of the red fermented beancurd and 1 tbsp of sugar from the marinade, mix with 1 cup of water, and stir until dissolved. Soak in the lotus root for around half an hour. Set aside.
2. Chop the chicken into pieces, rinse and drain. Mix with the marinade and leave for half an hour.
3. Heat some oil, coat the chicken with flour, deep-fry in hot oil until fully cooked and golden, set aside.
4. Drain the lotus root, sop up the moisture with a dry cloth, then deep-fry in hot oil until golden. Dish up and drain.
5. Heat some oil, sauté the spices until fragrant, put in the chicken and stir-fry. Finally sprinkle with the Hua Diao wine, give a good stir-fry and dish up. Put the deep-fried lotus root on the side and serve.

Liu's tips
· Choose lotus root that is long and thin body.
· The lotus root for deep-frying must be cut into thin slices (about 2 to 3 mm), because it is hard to make thick pieces crunchy. The temperature of oil should not be too high, making it around 120°C. Remove the lotus root when it is crisp.

Cookware:
Wok

魚香茄子煲

Eggplant with Pork and Salted Fish in Casserole

Ingredients
300 g eggplant
160 g minced pork
3 dried black mushrooms
1 small piece Mei Xiang
salted fish (about 20 g)
1 cup water

Spices
finely chopped garlic
diced ginger
diced spring onion

Marinade
1/3 tsp salt
2 tsp caltrop starch
1 tsp sesame oil

Seasoning
1/3 tsp salt
1/2 tsp sugar
2 tsp oyster sauce
1 tsp chilli bean sauce
1 tsp dark soy sauce
1 cup water

Method
1. Heat 2 tbsp of oil in a wok, fry the salted fish over low heat and finely dice. Set aside.
2. Mix the minced pork with the marinade. Heat the wok, put in 1 tbsp of oil and stir-fry the pork until cooked. Set aside.
3. Soak the dried black mushrooms in water until soft, remove the stalks and finely dice. Rinse the eggplant, peel alternately and cut into sections. Put half cup of oil into the wok, heat up, deep-fry the eggplant until tender and golden, drain well.
4. Heat some oil, fry the spices over medium heat until fragrant, add all the ingredients and give a good stir-fry over medium heat. Mix in the seasoning, turn to low heat and simmer until the eggplant is soft. Thicken the sauce with caltrop starch solution, bring to the boil, transfer to a casserole and serve.

Liu's tips
· If you are not ready to cook the peeled and sectioned eggplant, soak it in diluted salted water to prevent it from turning brown as a result of oxidation.
· Mei Xiang salted fish has a strong savoury smell. You can use it to give a special flavour to the dish.

Cookware:
Wok

Stir-fried Sea Cucumber with Zucchini and Bell Peppers

Ingredients
200 g frozen sea cucumber
 (rehydrated)
150 g zucchini
40 g dried wood ear fungus
100 g red, yellow and
green bell peppers

Ingredients for blanching
1 tsp salt
1 tsp sesame oil
1 tsp chicken bouillon powder
1 cup water
1 tsp cooking wine

Spices
finely chopped garlic
sliced ginger
sliced carrot

Thickening sauce
1/3 tsp salt
1 tbsp oyster sauce
1/2 tsp sesame oil
2 tsp caltrop starch
1 tbsp water

Method

1. Soak the frozen sea cucumber and wood ear fungus in water until soft. Put into hot water, add 2 tbsp of ginger juice wine, thoroughly blanch to remove the odd smell. Dish up and rinse in cold water.
2. Cut the sea cucumber and wood ear fungus into strips, add water and the ingredients for blanching. Cook for a while, dish up with the cooked sauce and set aside.
3. Rinse the zucchini and tri-coloured bell peppers and cut into strips. Heat a wok, add 1 tbsp of oil, stir-fry zucchini quickly over medium heat for keeping the zucchini crisp, add 1 tbsp of ginger juice wine and 1/2 tsp of salt, stir-fry for about 2 minutes. Put in the bell peppers and stir-fry again for about 1 minute and drain.
4. Drain the sea cucumber and wood ear fungus, set aside.
5. Heat the wok, add 1 tbsp of oil, sauté the spices until fragrant, put in all the ingredients and give a good stir-fry over medium heat. Pour in the mixed thickening sauce, stir-fry over high heat until hot. Serve.

Liu's tips

Method of preparing ginger juice wine: Blend 300 g of skinned ginger with 300 ml of cold boiled water in a food processer (the proportion of ginger to water is 1 : 1) and then sieve the juice. Add 50 ml of cooking wine into the juice and mix up. The ginger juice wine can help remove the fishy, odd smell of the food ingredients. It can be stored in a fridge for use anytime.

Cookware:
Wok

Minced Beef Rice with Dried Octopus and Water Chestnut

鱆魚馬蹄牛崧飯

Ingredients
250 g minced beef
6 skinned water chestnuts
5 dried black mushrooms
200 g dried octopus
20 g diced ginger
20 g diced spring onion
10 g finely chopped
dried tangerine peel

Marinade
1 tbsp light soy sauce
1 tbsp oyster sauce
3 tsp caltrop starch
4 tbsp water
1/4 tsp ground white pepper
1 tsp sesame oil

Mixed soy sauce
3 tbsp light soy sauce
2 tsp dark soy sauce
1 tsp sugar
1 tsp oil
1/2 tsp sesame oil
3 tbsp boiled water
ground white pepper

Method
1. Soak the dried octopus in water until soft, cut into dices and set aside.
2. Soak the dried black mushrooms in water until soft, remove the stalks and cut into cubes. Bash the water chestnuts and finely chop. Mix the minced beef with the marinade, add the black mushrooms, water chestnuts, dried tangerine peel and diced ginger, stir until gluey.
3. Heat the mixed soy sauce and set aside.
4. Rinse the white rice, put into a rice cooker with the dried octopus, add 300 ml of water and start cooking. When the water starts to dry after around 15 minutes, lay the minced beef flat on the rice, put the lid on and cook until the rice is done. Leave with the lid on for about 5 minutes.
5. Before serving, sprinkle with the diced spring onion, season with the mixed soy sauce and give a good stir. Serve.

Liu's tips
It is easier to dice the dried octopus after soaking. The beautiful smell of the octopus will infuse into the rice by cooking it with the white rice and water.

Cookware:
Rice cooker

Stir-fried Mixed Vegetables in Taro Nest

Ingredients
8 fresh black mushrooms
8 button mushrooms
8 fresh straw mushrooms
150 g celery
80 g carrot
120 g vegetarian meat
1 sandwich bread
300 g taro

Seasoning
1/3 tsp salt
1 tbsp oyster sauce
2 tsp caltrop starch
2 tbsp water

Method for taro nest
1. Thinly slice the taro, cut into fine shreds, mix with 2 tsp of salt and leave for about 8 minutes. Wash away the salt, drain and sop up the water with a dry cloth as far as possible.
2. Fill a wok halfway with oil and heat up over medium heat. Put in 2 bowls for moulding the nest until they are covered with the oil. Remove and set aside.
3. Mix the shredded taro with 2 tbsp of flour. Place a couple of taro shreds on the bottom of the bowl first. Use 3 to 4 taro shreds to form a circle, arrange the circles along the edge of the bowl in layers, and work into the shape of a rattan basket. Finally put a few taro shreds on the bottom to stabilize.
4. Put another bowl into the bowl containing the taro shreds. Press lightly, put into hot oil and then deep-fry over medium heat until the nest turns golden (or it comes off the mould). When the nest comes off, deep-fry for a moment and dish up.

Method

1. Crisscross the black mushrooms to 4 small pieces. Cut the button mushrooms and straw mushrooms in half. Blanch in ginger juice with wine, dish up and then rinse in cold water. Set aside.
2. Dice the vegetarian meat, celery and carrot. Set aside.
3. Remove the crust of the sandwich bread, cut into cubes and deep-fry in oil over medium heat. When they turn golden, dish up and sop up the oil with kitchen paper. Let cool.
4. Pour 1 cup of water into a wok, add 1/2 tsp of salt and 1 tsp of oil and bring to the boil. Put in the celery and carrot and blanch for about 1 minute. Add all the mushrooms and vegetarian meat, heat up. Drain and set aside.
5. Heat the wok and put in 1 tsp of oil. Add all the ingredients and give a good stir-fry. Pour in the mixed seasoning and stir-fry evenly. Place into the taro nest and sprinkle the crunchy bread on top. Serve.

Liu's tips

· You may bake the bread cubes instead. Put them on a baking tray and then bake at 120°C for about 5 minutes or until golden. Remove and let cool.
· Your hands will easily get itchy after handling the taro as its skin contains oxalic acid. It is better to remove the skin before washing it.
· Gently mix the taro shreds with salt to make them soft. It is easier to mould softened taro shreds into a nest.
· Before using the bowls for moulding the nest, sterilize them with boiling water and wipe them dry with a cloth. Put the bowls into hot oil over medium heat and the oil should cover the bowls. Remove the bowls and arrange the taro shreds on it. Otherwise, the taro shreds will stick to the bowl and the nest can hardly come out.

Cookware:
Wok and oven

Stir-fried Japanese Eel in Chilli Bean Sauce

豉椒炒鱔球

Ingredients
500 g boned Japanese eel
1/2 green bell pepper
1 red chilli
1/2 onion

Spices
1 tsp finely chopped garlic
1 tbsp fermented black bean sauce
(refer to p.8)
20 g diced spring onion

Marinade
1/2 tsp salt
3 tsp caltrop starch
ground white pepper
sesame oil
3 tsp water

Seasoning
1/3 tsp salt
1/2 tsp sugar
1 tbsp oyster sauce
2 tsp dark soy sauce
2 tsp caltrop starch
2 tbsp water
ground white pepper
sesame oil

Cookware:
Wok

Method
1. Rinse the Japanese eel. Score lengthwise and then across into a pattern. Cut into pieces of 3 cm x 6 cm long. Mix with the marinade, set aside.
2. Rinse the green bell pepper, red chilli and onion. Cut into wedges.
3. Heat a wok and deep-fry the Japanese eel over medium heat until fully cooked. Set aside.
4. Put the spices into the wok and sauté until fragrant. Add the green bell pepper, red chilli and onion, stir-fry over low heat until cooked. Turn to medium heat, put in the Japanese eel and stir-fry roughly. Put in the mixed seasoning and give a good stir-fry. Put on a plate and serve.

Liu's tips
You can ask the fishmonger to bone the Japanese eel for you, but the sticky skin mucus cannot be removed completely. Spread 2 tsp of salt on the skin and leave for about 1 minute; or soak the eel in warm water at about 75°C for 1 minute to let the white sticky skin mucus appear. Scrape it off with a knife and rinse the eel thoroughly.

Crab Satay with Mung Bean Vermicelli in Casserole

沙爹粉絲蟹煲

Ingredients
1 male mud crab
(about 600 g)
1 small bundle
mung bean vermicelli
100 g sliced ginger
80 g sectioned spring onion
100 g Chinese celery
80 ml coconut milk

Spices
1 tsp finely chopped garlic
1 tbsp diced red chilli

Seasoning
1 tsp salt
1 tsp sugar
1/2 tsp chicken bouillon
powder
2 tbsp satay sauce
1.5 cups water

Method
1. Soak the mung bean vermicelli in water until soft and drain. Heat 1 tbsp of satay sauce with 1/2 cup of water. Soak in the mung bean vermicelli and set aside.
2. Chop into the abdomen of the crab, open the crab shell, remove the internal organs, rinse and chop into pieces. Drain and mix with 1 tbsp of caltrop starch.
3. Heat 2 tbsp of oil and fry the crab over medium heat until fragrant. Add the ginger and spices, sauté until aromatic. Pour in the seasoning and heat up, put a lid on and then turn to low-medium heat to cook for around 6 minutes.
4. Remove the lid, mix in the spring onion and Chinese celery, turn to low heat and cook for about 2 minutes with the lid on. Turn off the heat, pour in the coconut milk and mix well.
5. Drain the mung bean vermicelli, put into a casserole and arrange the crab satay on top. Serve.

Liu's tips
The mung bean vermicelli will easily swell and appear too soft if it is cooked with the crab and seasoning. Just put it on the bottom of the casserole right after the crab is done to let it absorb the flavour and essence of the sauce.

Cookware:
Wok

177

荔茸海鮮卷

Seafood Rolls with Mashed Taro

Ingredients
12 frozen scallops
12 frozen shelled shrimps
10 spring roll pastry

Ingredients of mashed taro
300 g taro
40 g Tang flour
30 g vegetable shortening

Seasoning for mashed taro
1/2 tsp salt
1/2 tsp chicken bouillon powder
1/2 tsp five-spice powder
1/2 tsp sesame oil

Seasoning for seafood
1/2 tsp salt
1 tsp caltrop starch
1 tsp sesame oil

Batter
2 tbsp flour
4 tbsp water

Method
1. Cut the taro into pieces, steam until soft and crush into puree. Set aside.
2. Mix the Tang flour with 40 ml of boiling water, knead into half-cooked dough, mix with the mashed taro and seasoning. Finally add the vegetable shortening and knead until the dough is smooth.
3. Rinse the scallops and shrimps, sop up the water with a dry cloth, mix with the seasoning for seafood and leave for around half an hour. Cut into coarsely cubes, blanch in boiling water until fully cooked and drain. Add to the mashed taro and mix well as filling.
4. Mix the batter until it is sticky. Set aside.
5. Flatten the spring roll pastry, put in the taro filling and wrap into a rectangular-shaped spring roll. Seal the opening with the batter, put into hot oil and deep-fry over medium heat until it is crisp and golden. Serve.

Liu's tips
After the half-cooked dough is mixed with the mashed taro, let it cool down first; otherwise, the heat of the dough will make the vegetable shortening dissolve and the mashed taro will fall apart.

Cookware:
Wok

Deep-fried Almond Chicken in Lemon Sauce

杏香西檸雞

Ingredients
1 boned chicken thigh
200 g sliced almond
100 ml lemon sauce
 (refer to p.9)
4 slices lemon
2 tsp caltrop starch
 (mixed with water)

Marinade
1/2 tsp salt
1 egg yolk
2 tsp caltrop starch

Batter
1 egg (whisked)
4 tbsp caltrop starch

Method
1. Rinse and drain the chicken thigh. Mix with the marinade and rest for half an hour.
2. Mix the batter, pour into the chicken thigh and mix well. Roll the chicken thigh on the sliced almond.
3. Fill a wok halfway with oil, heat until the oil is medium hot (about 140°C), and then deep-fry the chicken thigh for around 1 minute. Turn to low-medium heat (about 100°C) and deep-fry for about 3 minutes, or until fully cooked. Finally adjust to medium heat and deep-fry the chicken thigh until the surface is crisp and golden. Drain the oil and cut into small pieces. Put on a plate.
4. Heat the lemon sauce and lemon slices. Thicken the sauce by stirring in 2 tsp of caltrop starch solution. Bring to the boil and pour onto the chicken thigh. Serve.

Liu's tips
The temperature of oil for deep-frying the chicken thigh should be medium hot. Too low the temperature will easily make the sliced almond fall off. If too high, the almond will burn and the chicken thigh will not be fully cooked.

Cookware:
Non-stick wok

茄汁煎中蝦

Fried Prawns in Sweet and Sour Sauce

Ingredients
12 fresh medium prawns
80 g onion

Spices
finely chopped garlic
sliced ginger
diced spring onion

Sauce
1/2 tsp salt
1/3 tsp chicken
bouillon powder
2 tsp sugar
3 tbsp ketchup
2 tbsp sweet and sour sauce
 (refer to p.7)
1 tsp caltrop starch

Method
1. Cut away the legs and tentacles of the medium prawns with scissors; rinse and drain. Coarsely shred the onion.
2. Heat a wok, put in 2 tbsp of oil, and fry the prawns over medium heat until both sides are golden and fully cooked. Set aside.
3. Put in the onion and stir-fry until fragrant. Add the spices and sauté until aromatic. Put in the prawns and give a good stir-fry over medium heat. Pour in the mixed sauce and stir-fry evenly over high heat. Serve.

Liu's tips
When removing the legs and tentacles of the prawn, you may cut its back open to pick out the intestine. It can also help the prawn to cook through and absorb the flavour of sauce easily.

Cookware:
Wok

Braised Dongpo Pork with Winter Melon

東坡白玉

Ingredients
600 g winter melon
600 g pork belly
80 g Chinese chives

Spices
120 g sliced ginger
100 g sectioned spring onion
5 star anise
1 whole piece dried
tangerine peel

Seasoning
1 tsp salt
2 tbsp oyster sauce
1 tbsp Chu Hou sauce
2 tbsp light soy sauce
1 tbsp dark soy sauce
1 tbsp sugar
100 ml Hua Diao wine
3 cups water

Method
1. Cut the winter melon into 4 cm cubes (about 2 cm thick), blanch in boiling water for about 5 minutes, rinse in cold water and set aside.
2. Cut away the layer of fat with skin on from the pork belly, leaving about 2 cm thick of the lean pork, and then rinse.
3. Heat the seasoning, pour into a rice cooker, add the spices and pork belly, cook for about 1.5 hours. Remove and let cool, cut into 4 cm cubes. Keep the sauce for later use.
4. Soak the Chinese chives in hot water until soft. Rinse in cold water and set aside.
5. Put the winter melon on top of the pork belly, tie with the Chinese chive in crisscross, and put on a deep dish. Cover with the remaining sauce, steam for about 15 minutes, or until the winter melon is tender, transfer to a plate. Use the sauce to make thickening glaze and pour over the winter melon. Serve.

Liu's tips
· The fat of the pork belly is too greasy. It will give a fresh taste by replacing the fat part with winter melon.
· The surface layer of the winter melon is firmer in texture and will not easily fall apart during cooking. When cutting the winter melon, keep that layer in 2 cm thick.

Cookware:
Wok and rice cooker

生財蒜子瑤柱甫

Steamed Dried Scallops and Deep-fried Garlic with Chinese Lettuce

Ingredients
12 whole dried scallops
80 g garlic cloves
300 g Chinese lettuce
1 slice ginger

Seasoning and thickening glaze
1 tbsp oyster sauce
1 tsp dark soy sauce
2 tbsp caltrop starch

Method
1. Put the whole dried scallops into a small deep dish. Cover with about 1 cup of water and soak for around 2 hours.
2. Skin the garlic cloves and rinse. Put 1 tbsp of oil into a wok and fry the garlic cloves over low heat until golden. Place the garlic cloves over the dried scallops, add the ginger and steam for about 1 hour.
3. Rinse the Chinese lettuce, blanch in boiling water until cooked, set aside.
4. Take out the small dish of dried scallops, gently pour out the steamed scallop water and keep the water. Place the Chinese lettuce into the dish of steamed scallops, slightly press, and firmly cover with a large deep dish. Turn the dish upside down, the small one on top and the large one on the bottom.
5. Pour the steamed scallop water into the wok, put in the seasoning and cook into a thickening glaze. Remove the small deep dish. Pour the thickening glaze over the steamed scallops. Serve.

Liu's tips
· It is better to pick the lumps off the edge of the dried scallops; otherwise, the steamed scallops will taste a bit chewy.
· The whole dried scallops must be soaked in cold water to swell before they are steamed. If not, the steamed scallops will not swell entirely and can hardly be tender.

Cookware:
Wok

Baked Pork Hock in Chinese Marinade

Ingredients
2 pork hocks

Spices
10 shallots
6 garlic cloves
4 slices ginger

Ingredients of
Chinese marinade
6 star anise
2/3 dried tangerine peel
8 slices sand ginger
8 bay leaves

Seasoning
2 red fermented beancurd
2 tsp salt
2 tsp chicken bouillon
powder
3 tbsp oyster sauce
30 g rock sugar
3 tbsp Hua Diao wine
2 tsp black peppercorns
2 tbsp Chu Hou sauce
8 cups water

Method
1. Chop the pork hocks, blanch in boiling water, rinse and put into a rice cooker with the ingredients of the Chinese marinade.
2. Heat a wok, put in 2 tbsp of oil, sauté the spices until fragrant, add the seasoning and heat up. Pour into the rice cooker and simmer for about 1 hour and 30 minutes. Turn off the rice cooker and rest for half an hour. Transfer the pork hocks onto a baking tray with the skin side up.
3. Preheat an oven to 240°C for 10 minutes, put in the pork hocks and bake for about 15 minutes. Put on a plate and serve.

Liu's tips
When the pork hocks are ready to be baked, wipe the moisture away from the skin and place them with the skin side up. It will help make the skin crunchy.

Cookware:
Wok, rice cooker and oven

Cheese Baked Tiger Prawns with Bacon

煙肉芝士焗大花蝦

Ingredients
6 tiger prawns
1 onion
3 slices bacon
2 tsp mixed herbs

Cheese sauce
4 slices Cheddar cheese
25 g butter
1 tsp sugar
1/3 tsp chicken bouillon powder
2 tsp flour
80 ml water

Method
1. Cheese sauce: Melt the butter over low heat, gently add the flour and mix well. Pour in the water and heat up. Add the other ingredients and cook until they dissolve. Set aside.
2. Dice the bacon. Cut the onion into round pieces and put on a baking tray.
3. Remove the shell from the body of the tiger prawns (keep the head and tail). Slit the middle of the body open, rinse and drain. Put the tail through the gap on the body, lightly pull straight and blanch in boiling water to shape.
4. Sop up the water from the tiger prawns and place over the onion in the baking tray. Pour the cheese sauce onto the prawns, and sprinkle with the bacon and herbs. Put into an oven which has been preheated for 10 minutes, and bake at 220℃ for about 5 to 6 minutes, or until the surface is golden. Serve.

Liu's tips
· The moisture on the blanched tiger prawns must be sopped up before adding the cheese sauce. If they are soggy, the cheese sauce will not be baked sticky.
· Do not bake the prawns for too long; or it will be tough.

Cookware:
Wok and oven

Deep-fried Fish in Sweet and Sour Sauce

糖醋菊花魚

Ingredients
600 g boned grass carp
150 ml sweet and sour sauce
(refer to p.7)

Spices
80 g red, yellow and
green bell pepper (diced)
1/2 tsp finely chopped garlic
1 tsp diced spring onion

Marinade
1 tsp salt
1/2 tsp chicken
 bouillon powder
2 tsp sesame oil
1/2 tsp ground white pepper
1 egg
3 tsp cornstarch (added last)

Method
1. Rinse the grass carp and wipe dry. Score the meat deep down to the skin (do not cut off) lengthwise and across into a pattern, and then cut into pieces.
2. Mix the grass carp with the marinade, leave for about 15 minutes and coat evenly with cornstarch.
3. Fill a wok halfway with oil, heat until it is medium hot (about 150℃), deep-fry the grass carp until golden, drain the oil. Put on a plate.
4. Heat the sweet and sour sauce, add the spices and give a stir-fry. Thicken the sauce with 2 tsp of caltrop starch solution, bring to the boil and pour over the grass carp. Serve.

Liu's tips
· When scoring the fish into a pattern, make it deep down to the skin. The deeper the cut is made, the clearer the pattern is shown on the deep-fried fish, looking like the shape of a chrysanthemum. You should be careful of not cutting the skin off.
· Coat every part of the fish with cornstarch evenly; otherwise it will not taste fluffy and crunchy.
· Shake off excessive cornstarch from the fish before it is deep-fried. It will help loosen the scored meat to give a clearer pattern after deep-fried.
· The temperature of oil for deep-frying the fish should not be too low; otherwise, the cornstarch will easily fall apart.

Cookware:
Non-stick wok

Deep-fried Sweet Potato Dumplings

薯茸黃金角

Ingredients
150 g glutinous rice flour
200 g Japanese sweet potatoes
80 g Tang flour
80 ml boiling water
20 g sugar
2 tsp custard powder
1 tsp oil
90 ml water

Filling
3 tbsp brown sugar
2 tbsp white sesame seeds
100 g crushed deep-fried peanuts
2 tbsp shredded coconut
2 tbsp peanut butter

Method
1. Mix the filling well and set aside.
2. Mix 80 ml of boiling water with 80 g of Tang flour, knead into cooked dough.
3. Steam the sweet potatoes until cooked, remove the skin and crush into puree.
4. Mix the glutinous rice flour with water, add the custard powder, sweet potato puree, sugar and cooked dough, and then knead into dough. Finally add the oil and knead evenly.
5. Divide the dough into small pieces weighting about 25 g each, press into a round shape, wrap 1 tsp of the filling in and shape into a triangle.
6. Heat oil in a wok. When the temperature of oil comes to about 120°C, put in the dumplings and deep-fry for about 3 minutes, or until they slightly float. Turn to medium heat and deep-fry until the surface is golden and crisp. Serve while hot.

Liu's tips
· The oil for deep-frying must cover the dumplings to provide enough room for the dumplings to float. It is also not easy for them to stick to the wok and burn.
· Heat the oil over medium heat and adjust to low heat to deep-fry the dumplings. Finally return the heat to medium and deep-fry the dumplings until the skin is crunchy.

Cookware:
Wok

Glutinous Rice Balls
Stuffed with Egg Custard

奶皇糯米糍

Ingredients of glutinous rice ball skin

150 g glutinous rice flour
40 g Tang flour
120 g sugar
200 ml coconut milk
100 ml milk
200 g shredded coconut

Ingredients of egg custard filling

250 ml milk
200 ml coconut milk
90 g sugar
25 g gelatin powder
40 g custard powder
30 g milk powder
40 g butter
120 ml water
1 egg (whisked)

Method

1. Mix together the ingredients of the glutinous rice ball skin (except the shredded coconut), pour into a dish and steam for about half an hour. Let cool and set aside.
2. Mix the sugar and gelatin powder of the ingredients of egg custard filling. Add water and mix well, pour into a dish and steam until they dissolve (about 10 minutes). Give a good stir and set this part of filling aside.
3. Steam the butter until it melts, mix in all the ingredients of the egg custard filling, sieve with the mesh and put into a dish, steam for about half an hour. Take out and let it cool, put into a fridge to chill for half an hour. Divide into small pieces for easy wrapping.
4. Knead the glutinous rice skin dough into an elongated shape, cut into small pieces and flatten. Wrap the egg custard filling in the skin, knead into small balls and coat with the shredded coconut. Serve.

Liu's tips

· The steamed egg custard filling should be mixed well, cooled down and chilled for about half an hour. It will become a bit firm for easy wrapping.
· If the glutinous rice skin dough sticks to your hands while kneading, you can sprinkle some shredded coconut on it and continue to knead.

Cookware:
Wok or large stainless steel stockpot

西米焗布甸

Baked Sago Pudding

Ingredients
70 g sago
160 g lotus seed puree
60 g sugar
20 g butter
40 ml evaporated milk
80 ml milk
2 egg yolks
30 g custard powder
300 g water

Method
1. Put the sago in boiling water, bring to the boil and turn off the heat. Put a lid on and leave until they are completely transparent. Dish up and cool in iced water or cold water.
2. Put 20 g of the lotus seed puree into each ramekin.
3. Heat the water; add the sugar, butter, evaporated milk, milk and egg yolks and heat up. Dissolve the custard powder in 40 ml of water, stir into the milk mixture to thicken it. Finally add the sago, mix well and put into the ramekins.
4. Preheat an oven for 15 minutes, arrange the ramekins into an oven, bake at 180°C for about 20 minutes, or until the surface is golden. Serve.

Liu's tips
· The sago will melt and will not become transparent by cooking in boiling water. It needs to be soaked in boiling water until it swells, and then soaked in cold water to become soft and transparent.
· If you use glass ramekins, put a little water into the baking tray, arrange the glass ramekins on the tray and start baking. It can avoid the ramekins cracking at the high temperature.

Cookware:
Oven

Mini Walnut Cookies

Ingredients
300 g low-gluten flour
100 g sugar
150 g shortening
60 g walnuts
1 egg
1 egg wash
40 ml water
2 g baking soda
4 g baking powder

Method
1. Sieve the low-gluten flour with a fine meshed sieve, set aside.
2. Mix the sugar, shortening, baking soda and baking powder together until the sugar dissolves. Add 1 egg and the water, then knead evenly. Gradually add the flour and knead until the dough is smooth, cover with the cling wrap and leave for 20 minutes.
3. Knead the dough into an elongated shape, cut into small pieces, knead into a round shape and put on a baking tray.
4. Gently press the small dough, make a tiny well in the middle, put in the walnuts, and brush the egg wash on the entire surface.
5. Set the temperature of an oven at 180°C and preheat for 15 minutes. Bake the walnut cookies with upper heat at 180°C and lower heat 100°C for about 15 minutes, or until they turn slightly brown. Take out and brush the surface with the egg wash. Adjust the upper heat to 150°C, turn off the lower heat, and bake for 5 minutes. Turn off the heat, leave for 3 minutes. Serve.

Liu's tips
· It is not appropriate to knead the dough heavily. Knead it gently; otherwise, it will produce a higher content of gluten, the baked cookies will taste hard and will not be fluffy and crisp.
· If your oven is not equipped with individual settings for the temperature of upper and lower heat, then bake the cookies with both upper and lower heat at 160°C for 15 minutes. Take out the cookies, brush the surface with egg wash, adjust the upper heat to 150°C, turn off the lower heat and bake for 5 minutes. Turn off the oven and leave for 3 minutes.
· Do not arrange the cookies on the baking tray too closely. Reserve some space in between to let them prove and swell.
· Let the baked walnut cookies cool down completely before eating. The residual heat will help release the smell of the baking powder.

Cookware:
Oven

蓮子綠豆爽

Lotus Seed and Mung Bean Sweet Soup

Ingredients
200 g shelled mung beans
120 g lotus seeds
120 g skinned water chestnuts
2 slices ginger
2 eggs (whisked)
6 cups water
25 g water chestnut flour

Seasoning
160 g rock sugar

Method
1. Soak the lotus seeds and shelled mung beans in water separately. When the lotus seeds turn soft, remove the cores. Blanch the mung beans in boiling water for about 15 minutes, drain and set aside. Rinse the skinned water chestnuts and cut into thin slices.
2. Pour water into a saucepot, add the ginger and lotus seeds, cook for about 45 minutes, or until the lotus seeds are tender. Add the mung beans and water chestnuts and bring to the boil. Season with the rock sugar.
3. Dissolve the water chestnut flour in some water, pour into the sweet soup and stir to thicken the soup. Bring to the boil, turn off the heat, add the egg wash bit by bit while stirring the sweet soup gently. Serve.

Liu's tips
· The lotus seeds can hardly be cooked tender with sugar. Add the rock sugar only after the lotus seeds are cooked and become completely soft.
· Stir the egg wash in the sweet soup only after the heat is turned off; otherwise the egg wash will be overcooked and will not be smooth.

Cookware:
Saucepot

Water Chestnut Cake with Sugarcane and Couchgrass Root Flavour

竹蔗茅根馬蹄糕

Ingredients

150 g skinned water chestnuts
120 g water chestnut flour
3 g custard powder
100 g palm sugar
200 ml water
500 ml sugarcane and
 couchgrass root drink
(sold in supermarkets in bottles)

Method

1. Rinse the skinned water chestnuts, thinly slice and set aside.
2. Put the water chestnut flour, custard powder and water into a dish and stir until dissolve.
3. Bring the sugarcane and couchgrass root drink with the palm sugar to the boil, add the sliced water chestnuts and continue to cook.
4. Brush a cake mould thinly with oil. Pour the warm sugarcane and couchgrass root drink into the flour mixture, stir until it becomes batter. Pour into the cake mould and steam for about 30 minutes, let cool and cut into pieces. Serve.

Liu's tips

· Pay attention to the temperature of the boiled sugarcane and couchgrass root drink when stirring it into the flour mixture. If it is too hot, the flour mixture will be overcooked and turn thick. It will give an uneven surface after pouring into the cake mould, making the presentation less beautiful.
· You can serve the steamed water chestnut cake cold or hot. If you want it hot, cut it into pieces and fry until fragrant; or steam the pieces of cake again after removing it from the fridge.

Cookware:
Wok or large stainless steel stockpot

4廚具煮好菜 Home Favourites with 4 Cookwares

作者	Author
廖教賢	Alvin Liu
策劃/編輯	Project Editor
	Karen Kan
攝影	Photographer
	Imagine Union
美術統籌及設計	Art Direction
	Amelia Loh
美術設計	Design
	Man Lo
出版者	Publisher
	Forms Kitchen
	an imprint of Forms Publications (HK) Co. Ltd.
香港英皇道499號北角工業大廈18樓	18/F, North Point Industrial Building, 499 King's Road, Hong Kong
電話	Tel: 2138 7998
傳真	Fax: 2597 4003
網址	Web Site: http://www.formspub.com
	http://www.facebook.com/formspub
電郵	Email: marketing@formspub.com
發行者	Distributor
香港聯合書刊物流有限公司	SUP Publishing Logistics (HK) Ltd.
香港新界大埔汀麗路36號	3/F., C&C Building, 36 Ting Lai Road,
中華商務印刷大廈3字樓	Tai Po, N.T., Hong Kong
電話	Tel: 2150 2100
傳真	Fax: 2407 3062
電郵	Email: info@suplogistics.com.hk
承印者	Printer
中華商務彩色印刷有限公司	C & C Offset Printing Co., Ltd.
出版日期	Publishing Date
二〇一五年七月第一次印刷	First print in July 2015

瀏覽網站

會員申請